T0236302

Lecture Notes in Artificial Intelligence   10089

Subseries of Lecture Notes in Computer Science

More information about this series at http://www.springer.com/series/1244

Rajendra Prasath · Alexander Gelbukh (Eds.)

Mining Intelligence and Knowledge Exploration

4th International Conference, MIKE 2016
Mexico City, Mexico, November 13–19, 2016
Revised Selected Papers

 Springer

Editors
Rajendra Prasath
Norwegian University of Science
  and Technology
Trondheim
Norway

Alexander Gelbukh
Center for Computing Research, CIC
National Polytechnic Institute, IPN
Mexico City, Distrito Federal
Mexico

ISSN 0302-9743   ISSN 1611-3349   (electronic)
Lecture Notes in Artificial Intelligence
ISBN 978-3-319-58129-3   ISBN 978-3-319-58130-9   (eBook)
DOI 10.1007/978-3-319-58130-9

Library of Congress Control Number: 2017938635

LNCS Sublibrary: SL7 – Artificial Intelligence

Printed on acid-free paper

This Springer imprint is published by Springer Nature
The registered company is Springer International Publishing AG
The registered company address is: Gewerbestrasse 11, 6330 Cham, Switzerland

Preface

This volume contains the papers presented at MIKE 2016: the 4th International Conference on Mining Intelligence and Knowledge Exploration held during November 13–19, 2016, at the Center for Computing Research (CIC) of the National Polytechnic Institute (IPN), Mexico City (http://www.mike.org.in/2016/). There were 56 submissions from 13 countries and each qualified submission was reviewed by a minimum of two Program Committee members using the criteria of relevance, originality, technical quality, and presentation. The committee accepted 17 full papers for oral presentation at the conference. The overall acceptance rate is 30.36%.

The International Conference on Mining Intelligence and Knowledge Exploration (MIKE) is an initiative focusing on research and applications on various topics of human intelligence mining and knowledge discovery. Human intelligence has evolved steadily over several generations, and today human expertise is excelling in multiple domains and in knowledge-acquiring artifacts. The primary goal was to focus on the frontiers of human intelligence mining toward building a body of knowledge in this key domain. The focus was also to present state-of-art scientific results, to disseminate modern technologies, and to promote collaborative research in mining intelligence and knowledge exploration. At MIKE 2016, specific focus was placed on the "learning to explore knowledge discovery" theme.

The accepted papers were chosen on the basis of their research excellence, which provides a body of literature for researchers involved in exploring, developing, and validating learning algorithms and knowledge-discovery techniques. Accepted papers were grouped into various subtopics including information retrieval, machine learning, pattern recognition, knowledge discovery, classification, clustering, image processing, network security, speech processing, natural language processing, language, cognition and computation, fuzzy sets, and business intelligence. Researchers presented their work and had an excellent opportunity to interact with eminent professors and scholars in their area of research. All participants benefitted from discussions that facilitated the emergence of new ideas and approaches. The authors of short papers presented their work during a special session and obtained feedback from thought leaders in the discipline.

A large number of eminent professors, well-known scholars, industry leaders, and young researchers participated in making MIKE 2016 a great success. We express our sincere thanks to the National Polytechnic Institute (IPN), Mexico City, Mexico, for allowing us to host MIKE 2016. We were pleased to have the following serving as advisory members for MIKE 2016: Prof. Ramon Lopaz de Mantaras, Artificial Intelligence Research Institute, Spain; Prof. Mandar Mitra, Indian Statistical Institute (ISI), Kolkata, India; Prof. Pinar Ozturk and Prof. Bjorn Gamback, Norwegian University of Science and Technology, Norway; Prof. Sudeshna Sarkar and Prof. Niloy Ganguly, Indian Institute of Technology, Kharagpur, India, Prof. Philip O'Reilly (University College Cork, Ireland); Prof. Nirmalie Wiratunga (Robert Gordan University, UK);

Prof. Paolo Rosso (Universitat Politecnica de Valencia, Spain); Prof. Chaman L. Sabharwal, Missouri University of Science and Technology, USA; and Dr. Rajarshi Pal (IRDBT, Hyderabad).

Several eminent scholars, including Prof. Ildar Batyrshin, National Polytechnic Institute, Mexico, Prof. Genoveva Vargas-Solar, CNRS, France, and Prof. Grigori Sidorov, Natural Language and Text Processing Laboratory, CIC - IPN, Mexico City, Mexico, delivered invited talks on learning and knowledge exploration topics in various interdisciplinary areas of artificial intelligence and machine learning.

We thank the Technical Program Committee members and all reviewers/subreviewers for their timely and thorough participation in the reviewing process.

We express our sincere gratitude to Prof. Marco Antonio Ramirez Salinas, Director of CIC-IPN, Mexico City, for his support in organizing MIKE 2016 in IPN, Mexico City. We also thank SMIA, the Mexican Society of Artificial Intelligence, for their support. We appreciate the time and effort put in by the members of the local organizing team at CIC-IPN, Mexico City. We are very grateful to all our sponsors, especially the Government of Mexico and many other local supporters, for their generous support to MIKE 2016.

Finally, we acknowledge the support of EasyChair in the submission, review, and proceedings creation processes. We are very pleased to express our sincere thanks to Springer, especially Alfred Hofmann, Anna Kramer, and the editorial staff, for their faith and support in publishing the proceedings of MIKE 2016.

November 2016   Rajendra Prasath
                                                                Alexander Gelbukh

Organization

Program Committee

Agnar Aamodt	Norwegian University of Science and Technology, Norway
Ibrahim Adeyanju	Federal University Oye-Ekiti, Nigeria
Rajendra Akerkar	Western Norway Research Institute, Norway
Gethsiyal Augasta	Sarah Tucker College (Autonomous), India
Zeyar Aung	Masdar Institute of Science and Technology, UAE
Rakesh Balabantaray	International Institute of Information Technology, Bhubaneswar, India
Lavanya Balaraja	University of Madras, Chennai, India
Anupam Basu	Indian Institute of Technology, Kharagpur, India
Pinaki Bhaskar	Samsung R&D Institute India Bangalore, India
Vasudha Bhatnagar	University of Delhi, New Delhi, India
Plaban Kumar Bhowmik	Indian Institute of Technology, Kharagpur, India
Isis Bonet	Universidad EIA, Envigado, Colombia
Erik Cambria	Nanyang Technological University, Singapore
Tanmoy Chakraborty	University of Maryland, College Park, USA
Joydeep Chandra	Indian Institute of Technology, Patna, India
Sanjay Chatterji	Samsung R&D Institute India, Bangalore, India
Manoj Chinnakotla	Microsoft R&D India Pvt. Ltd, India
Kamal Kumar Choudhary	Indian Institute of Technology, Ropar, India
Gladis Christopher	Presidency College, Chennai, India
Amélie Cordier	LIRIS, Université Claude Bernard Lyon 1, France
Amitava Das	Indian Institute of Information Technology, Sri City, India
Dipankar Das	Jadavpur University, India
Tirthankar Dasgupta	TCS Innovation Lab, New Delhi, India
Ramon Lopez De Mantaras	IIIA - CSIC, Barcelona, Spain
Maunendra Sankar Desarkar	Indian Institute of Technology, Hyderabad, India
Lipika Dey	TCS Innovation Labs, New Delhi, India
Aidan Duane	Waterford Institute of Technology (WIT), Ireland
Björn Gambäck	Norwegian University of Science and Technology, Norway
Vineet Gandhi	International Institute of Information Technology, Hyderabad, India
Debasis Ganguly	Dublin City University, Ireland
Niloy Ganguly	Indian Institute of Technology Kharagpur, India
Alexander Gelbukh	Instituto Politécnico Nacional

Bibudhendu Pati	C.V. Raman College of Engineering, Bhubaneswar, India
Soma Paul	International Institute of Information Technology, Hyderabad, India
Carla Pires	IFSUL, Brazil
Rajendra Prasath	Norwegian University of Science and Technology, Norway
Shantha Selvakumari R.	Mepco Schlenk Engg. College, Sivakasi, India
K. Sreenivasa Rao	Indian Institute of Technology, Kharagpur, India
Pattabhi Rk Rao	AU-KBC Research Centre, MIT Campus of Anna University, Chennai, India
Juan Recio-Garcia	University Complutense of Madrid, Spain
N. Subba Reddy	Gyeongsang National University, Jinju, South Korea
Paolo Rosso	Technical University of Valencia, Spain
Sudip Roy	Indian Institute of Technology Roorkee, India
Udayabaskaran S.	Vel Tech Technical University, Chennai, India
Chaman Lal Sabharwal	Missouri University of Science and Technology, USA
Moumita Saha	Indian Institute of Science, Bengaluru, India
Sudipta Saha	Indian Institute of Technology, Bhubaneswar, India
Sujan Kumar Saha	Birla Institute of Technology, Mesra, India
Saurav Sahay	Intel Labs, Santa Clara, CA, USA
Debasis Samanta	Indian Institute of Technology, Kharagpur, India
A.K. Sao	Indian Institute of Technology, Mandi, India
Sudeshna Sarkar	Indian Institute of Technology, Kharagpur, India
P. Shanmugavadivu	Gandhigram Rural Institute, Deemed University, India
Avinash Sharma	International Institute of Information Technology, Hyderabad, India
Dipti Misra Sharma	International Institute of Information Technology, Hyderabad, India
Manish Shrivastava	International Institute of Information Technology, Hyderabad, India
Manjira Sinha	Xerox Research Center India, India
Tripti Swarnkar	Indian Institute of Technology, Kharagpur, India
Jaisingh T.	Indian School of Mines, Dhanbad, India
Kathirvalavakumar T.	VHNSN College (Autonomous), Virudhunagar, India
Geetha T.V.	Anna University, College of Engineering, Chennai, India
Venu Thangaraj	Ramanujan Institute for Advanced Study in Mathematics, India
Diana Trandabat	University Al. I. Cuza of Iasi, Romania
Vasudeva Varma	International Institute of Information Technology, Hyderabad, India
Anil Kumar Vuppala	International Institute of Information Technology, Hyderabad, India
Nirmalae Wiratunga	Robert Gordan University, Aberdeen, UK
Wei Lee Woon	Masdar Institute, UAE

Additional Reviewers

Aggarwal, Apeksha
Alluri, Knrk Raju
Balaraja, Lavanya
Christopher, Gladis
Krishna, Hari
Lai, Mirko
Mondal, Anupam

Patra, Braja Gopal
Sarkar, Sandip
T. Kathirvalavakumar
T. Kumaran
Thirumuru, Ramakrishna
Vegesna, Vishnu Vidyadhara Raju

Contents

An Efficient Incremental Mining Algorithm for Dynamic Databases

Lydia Nahla Driff[(✉)] and Habiba Drias

Artificial Intelligence Laboratory (LRIA), Department of Computer Science,
USTHB, Bab Ezzouar, Algeria
driff.nahla@gmail.com, hdrias@usthb.dz

Abstract. Data mining is aimed to extract hidden acknowledge from large dataset, in order to exploit it for predicting future trends and make decisions. Extracting meaningful and useful candidate optimally is handled by several algorithms, mainly those based on exploring incoming data, which can lose information. To address this issue, this paper proposes an algorithm named Incremental Apriori (IncA) for discovering frequent itemsets in transaction databases, which is in fact a variant of the well-known Apriori algorithm. In IncA, we introduce a notion of promising items generated from the original database, an incremental technique applied on incremental database and a health check process to ensure candidate generation completeness. On the theoretical side, our algorithm exhibits the best computational complexity compared to the recent state-of-the-art algorithms. On the other hand, we tested the proposed approach on large synthetic databases. The obtained results prove that IncA reduces the running time as well as the search space and also show that our algorithm performs better than the Apriori algorithm.

Keywords: Datamining · Dynamic database · Apriori · Incremental technique · Machine Learning Techniques

1 Introduction

With the variation of communication means and popularization of information in diverse fields, information is growing rapidly databases size increases in the few past years from some Gigabytes to a thousand Exabyte [2]. This evolution gave birth to new challenges including analysis, data curation, sharing, storage, transfer [1] and search represent a big issue nowadays.

To extract useful information, data mining techniques propose diverse approaches. Most of them have to scan datasets to get more frequent information, as an example Apriori algorithm is designed to extract frequent pattern mining. Nowadays, the majority of databases are dynamic and are fed by new records or decremented by others. Consequently, old frequent items can become infrequent and new ones can be added. So determining the correct frequent itemsets to get the strong association rules and discard weak ones is a great challenge.

Therefore, this process requires a huge physical resource and large amount of computation time. Many incremental approaches have been developed to optimize

© Springer International Publishing AG 2017
R. Prasath and A. Gelbukh (Eds.): MIKE 2016, LNAI 10089, pp. 1–12, 2017.
DOI: 10.1007/978-3-319-58130-9_1

processing time by reducing the number of scan steps. However missing some frequent patterns is the great weakness of incremental approaches.

In this work, we propose a new approach called Incremental Apriori (IncA) which is an improved version of classic Apriori using clustering rules and Machine Learning Techniques. This approach make up the drawbacks of the existing approaches, first it assures reliability just by exploiting new flux, second it warrants completeness by founding all frequent patterns and keeping promoted patterns which can be frequent in further execution, all offering a minimum execution time and unheeded resort to old data.

2 Related Works

Discovering useful information from a large dataset needs a process with two major steps: generate frequent information and then find association rules. The first step is essential to insure the second one, thus stimulating the proposal of diverse algorithms. Apriori operates on databases containing transactions (for example, collections of items bought by customers, or details of a website frequentation) [12]. It is a classic algorithm for learning association rules by generating frequent candidate itemset (item combinations) and scanning database repeatedly starting from the size one-itemsets. Since the Apriori algorithm was proposed by Rakesh Agrawal [4], several studies have been conducted to improve it. For instance, the Direct Hashing and Pruning algorithm (*DHP*) [10] proceeds also to construct candidate itemsets through a set of iterations; and Frequent Pattern Growth (*FP-Growth*) [21], which stores a compressed format of frequent patterns information in the form of tree named FP-tree. Note that Apriori, DHP and FP-Growth are the three most popular algorithms especially for static databases.

Contrariwise, for dynamic database, several updating techniques have been developed. They are based on these two approaches, like Fast Update approach FUP, FUP2 and NewFUP, which proceed to update the obtained frequent itemset by finding remaining winners, removing losers and pruning candidate sets through scanning incremental database. FUTP2 is a complementary algorithm of FUP for finding new frequent itemsets in incremental database [13]. Therefore, FUP is very adaptive to size increase and can be applied to very large databases [11] but it can easily lose useful itemsets in pruning phase.

EDUA is an algorithm adopting the technique of scan reduction [7] in order to deal with multiple scans problem. It proceeds to store 2-itemsets candidate from the prior frequency in original database, generate k-itemsets candidate using scan reduction technique [14] and find frequent k-itemsets. In the case of incrementing, EDUA updates k-itemsets frequencies by scanning updated database, updates related 2-itemsets candidate and find new frequent itemsets in updated database. Then EDUA considers that the set of k-itemsets candidate is not much larger than the set of frequent k-itemsets, which is not a right assumption in most existing databases.

CanTree is an ameliorate incremental mining version of FP-Growth, it explore items proprieties, arrange them according to a canonical order and construct a tree format which will be visited for updated data [3]. This approach is independent of the threshold values, so it can require a big storage space.

Other algorithms follow different trends. They proceed to keep a maximum of items to ensure minimum loss of information, like a probability based incremental association rule discovery (probability based IARD) [15]. The latter handles new data, inserts into or deletes from a dynamic database using the principle of Bernoulli trials to select approximate frequent itemset and the principle of maximum possible value to define support counts. The weakness of this approach is that is based on probabilities to find bounders and it supposes that incremental database size is already known, which is not obvious. We can mention also the Three-Way Decision approach [16, 17], which is based on rules in probabilistic rough set models, which classify items in three regions: a positive region contains accepted items, a negative region contains rejected items and boundary region contains delayed items. The limits of those regions are defined using Bayesian decision procedure suggests the minimum-risk decision rule [18]. We can agree that Three-Way Decision approach explains rules also in probabilistic models.

Unlike keeping a maximum of items using probabilistic approach, BitApriori algorithm [8] performs the Bitwise "And" operation on the binary strings to get support count and built Trie data structure. The great disadvantage of this method is the memory high consumption. An improved version of this algorithm is proposed and called Modified BitApriori Algorithm [9]; those approaches are inspired from BitTableFI Algorithm [19] that stores a compressed format of database in order to generate candidate itemsets in special data structure called BitTable. It follows the same principle but performs the Bitwise "OR" operation in order to reduce the height of the Trie and insert only frequent items. Therefore, memory consumption is reduced, however Modified BitApriori Algorithm discards a large number of items.

Theoretical complexities of the different algorithms have not been compared, for several reasons, above all, those complexities are identical in the worst case. A study of J.L. Han and A.W. Plank [6] can serve as reference. On the other hand, this complexity is completely dominated by disk access cost [5].

3 Incremental Frequent Pattern Mining

As introduced previously, Incremental Apriori (IncA) is an enhanced version of Apriori, using previous mining to optimize new frequent patterns discovering.

3.1 Apriori Algorithm

Apriori is one of the most popular algorithm proposed by R. Agrawal and R. Srikant in 1994 for mining frequent item-sets and generating association rules [3]. In statistical databases, Apriori is the most suitable approach, because it uses an iterative approach called *level-wise* search for getting frequent itemsets. So, a whole database is scanned to get k-itemsets which are used to get (k + 1)-itemsets called potential large-itemsets by joining and pruning operations.

Minimum support *minSup* is a focal notion in this approach. It is calculated by fixing a coverage rate *cR* of database. An itemset is judged frequent if its frequency is superior/equal to minSup, Combination of frequent itemsets is extended to a new itemset called *candidate generation* [4].

Let *t* be an itemset and *F* a set of itemset. The Apriori process is as follows:

Apriori Algorithm

```
1) Fixe a coverage rate cR & calculate minSup
2) Forall (t ∈ database)
3)            freq(t)++
4) F1 ← t: freq(x) >= minSup
5) For (k=2; F(k-1)<>0; k++) {
6)            Ck← Join (F(k-1))
7)            Ck← Pruning (Ck)
8)            Forall (t ∈ Ck)
9)                       freq(t)++
10)           Fk ← t ∈Ck: freq(t) >= minSup
11)           F ← F∪Fk}
```

Where: *k* is the number of iterations.

The algorithm terminates when no further successful candidates are found by joining $F_{(k-1)}$ with itself.

Join function is described as follows:

```
Forall (t ∈ F(k-1))
         Forall (x ∈ F(k-1) & x !=t)
                  Ck← t ∪ x
```

Pruning function is described as follows:

```
Forall (c ∈ Ck)
            Forall (x ∈ Ck & x !=c)
                  If (c=x) delete x
Forall (c ∈ Ck)
            Forall (t ∈ F(k-1))
                  If (c=t) delete c
```

When analyzing Apriori, we can remark that the output contains only frequent itemsets, each itemset (including *k*-itemset) less than minSup is lost and no way to get a trace in the future. As well, the entire database is scanned for every new *k*-itemset. Let us calculate the temporal complexity of Apriori algorithm [6]. Note:

$|D|$ cardinality of our database and T number of $t \in$ database
$|F_k|$ & $|C_k|$ respective size of F_k & C_k and $|t|$ itemset size.
From Apriori algorithm above we have the following complexity:

Ins(2) →determinate F1 contains *1*-itemset according to minSup : $O(|D|*|T|)$

Inst(5) →repeat a set of sub-instructions k time, till no further successful candidate are found : $O(\Sigma\ k)$

Inst(6) →join $F_{(k-1)}$ to get candidate itemset : $O(|F_{(k-1)}|^2)$

Inst(7) →pruning C_k to eliminate repeated generated candidate itemset according to C_k and $F_{(k-1)}$: $O(\Sigma(|C_k|^2, |C_k|*|F_{(k-1)}|))$

Inst(8) →determinate F_k containing k-itemset : $O(|C_k|*|D|*|t|)$

So, final complexity is: $O'(|D| * |T| + \Sigma(|F_{(k-1)}|^2, |C_k|^2, |C_k| * |F_{(k-1)}|, |C_k| * |D| * |t|))$
In the worst case, F_k contains all items t ($|F_{(k-1)}| = |T|$) and C_k contains also all junction from $F_{(k-1)}$ ($|C_k| = 2^\wedge|F_{(k-1)}|$), so the theoretical complexity will be huge as well as spatial complexity.

3.2 Incremental Apriori Algorithm

Unlike the precedent issue, in the case of dynamic databases, the Apriori algorithm is not a suitable approach for insertion and deletion operations. But instead of reconsidering the whole new database, we search for techniques to use the results of the previous database with the introduced transactions. This task is called incremental mining.

The aim of the incremental approach is to determine all frequent itemsets just by exploring new incoming transactions. When a database is updated, some frequent itemsets still remain frequent or become infrequent, this is an easy case to check. However two cases are tricky to handle: when some infrequent itemsets are frequent in the updated database, or when some itemsets are infrequent in both databases. How can the system detect them as both are already discarded by the previous mining treatments? We resolve this problem by introducing the notion of promising items, which will be explained in the next sections. Likewise, the proposed algorithm minimizes also the disk access during the check of existing frequent itemsets with new incoming DB using an incremental approach.

Incremental Mining

As mentioned previously, checking existing frequent itemsets in a new incoming DB (noted uDB) is an easy task, but it takes a large number of disk accesses. A classic process proceeds to check each itemset and calculate its frequency in uDB. If the latter is equal or greater than the minimum support of the new DB namely $minSup_n$, it must be kept as frequent itemset, else it is discarded.

To overcome the time constraint, we propose an incremental approach, which is much rapid. It is outlined as follows:

Calculate the minimum support of uDB noted $minSup_u$, as: $minSup_u = |u\text{DB}| * cR/100$

- **step1.** get new frequent 1-itemset: For each item in uDB, increment its frequency and keep it if its frequency is equal or greater than $minSup_u$. From obtained frequent *1*-itemset in the last occurrence, only keep *1*-itemset existing in new getting frequent items (from uDB) and sum frequencies.
- **step2.** get new frequent k-itemset: For k-itemset where $k > = 2$, start from obtained frequent k-itemset in the last occurrence. Each time the itemset exists in uDB increment its frequency and keep only k-itemset that incremented counter is equal or greater than $minSup_u$. Incrementing frequency is doing parallel for each item (contained in k-itemset) in order to minimize execution time. Then, we cross new frequent *1*-itemset with rest of frequent k-itemset to obtain new k-itemset.

Incremental Apriori Algorithm

```
1)  minSup₁= |uDB| * cR /100
2)  Forall (t ∈ Fₖ)
3)       if (t=1-itemset) && (t ∈ uDB) {
4)              calculateFreq (t, uDB)
5)                     if (Freq(t) >= minSupᵤ) → addTo(Fₙ) }
         Else {
6)              Parallel { calculateFreq (tᵢ ∈ t, uDB) }
7)              ifall (Freq(tᵢ) >= minSupᵤ) → addTo(Fₙ)}
8)  Forall (t ∈ uDB)&&(t ∉ Fₖ){
9)         calculateFreq (t, uDB)
10)              if (Freq(t) >= minSupᵤ) → addTo(PFₙ)}
11) Cₙ← Join (Fₙ)
12) Fₙ← Pruning (Cₙ)
13) Forall (t ∈ Fₙ){
14)        calculateFreq (t, nDB)
15)              if (Freq(t) >= minSupₙ) → addTo(Fₙ)
```

Where: t is an itemset, ti is an item, Fk actual set of itemset, Fn new set of itemset, PFn kind of set of itemset (will be explained with more detail in the next section), Freq (t) is support of t in uDB, addTo() is a function used to add t to F_n

So, frequency of new frequent itemset is: $Freq_n(t) = Freq_o(t) + Freq_u(t)$

Time complexity

Let us calculate the temporal complexity of IncA algorithm. Note:

$|D|$ & $|D|$ | respective cardinality of updated database uDB & nDB
$|F_k|$, $|F_n|$ & $|C_n|$ respective size of F_k & C_n.
$|T'|$ number of t e uDB and ne F_k and $|t|$ itemset size.
From IncA algorithm in *Fig.4* we have the following complexity:
Ins(2) →check *1*-itemset & *k*-itemset according to $minSup_u$: $O(|F_n|)$
Inst(4) →calculate support of t (*1*-itemset) in uDB : $O(|D|)$
Inst(6) →calculate support of t (*k*-itemset) in uDB : $O(|D|*|t|)$
$\quad\quad\quad\quad$ → $O(\prod_1(|F_k|,\sum(|D|, |D|*|t|))$
Inst(8) →determinate new *1*-itemset according to $minSup_u$: $O(|D|*|T'|)$
Inst(9) →calculate support of new t (*1*-itemset) in uDB : $O(|D|)$
$\quad\quad\quad\quad$ → $O(\prod_2(|D|,|D|*|T'|))$
Inst(11) →join F_k & F_n to get candidate itemset : $O(|F_k|*|F_n|)$
Inst(12) →pruning C_n to eliminate repeated generated candidate itemset according to F_n: $O(|C_n|^2)$
Inst(13) →determinate F_k containing according to $minSup_n$: $O(|C_k|*|D'|*|t|)$

So, final complexity is: $O'(\Sigma(\prod_1, \prod_2, |F_k| * |F_n|, |C_n|^2, |C_k| * |D'| * |t|))$

In the worst case, all t e F_k exist in uDB ($|F_k| * |D|$ is maximized), and all generated itemsets from joining are new ($|C_k| * |D'|$ is maximized).

In two precedent cases, theoretical complexity is very less important than classic Apriori complexity, because it's mainly based on frequent itemsets size $|F|$ and updated database size $|D|$ not on the hole new database ($|F| << |D_n|$ and $|D| < |D_n|$), where $|D_n|$ is new database size.

Promising Items

Let oDB be the original database. By applying Apriori algorithm, we can found the set of frequent itemsets denoted F respecting the minimum support *minSup*. Let uDB be the inserted database and nDB the new updated database. $|n$DB$| = |o$DB$| + |u$DB$|$

A promising item is an infrequent item in oDB but suspected to be frequent in the future, its frequency is clearly less than *minSup*. The question that arises is: How can we judge that an item is promising?

It seems impossible to determinate uDB size before this base is added. So, we fixe uDB size by determinate a rate according to oDB size, called *Overlay* rate and denoted *ovR*, in order to assure that uDB size will be less than oDB size and no potential frequent itemsets will be lost. To select promising item we calculate potential bounder denoted $minSup_u$ using *ov* rate as follows:

$minSup_u = \max|u$DB$| * cR/100$ where: $\max|u$DB$| = |o$DB$| * ovR/100$

In the case of new incoming DB, Incremental approach (explain in previous section) is also apply on existing promising items in order to minimize execution time. Note that promising item is *1*-itemset. If a promising item becomes frequent, it will be join with existing frequent itemset in order to minimize storage space.

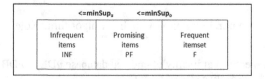

Fig. 1. Itemset categorization

Categorizing items and itemsets

We can conclude that an itemset can be in one of three distinguished sets. Figure 1 presents the categorization of oDB itemset according to the last example.

```
If (freq(t)>= minSup_o ) → addTo(F)
ElseIf (freq(t)< minSup_o & freq(t)>= minSup_u ) → addTo(PF)
ElseIf (freq(t) <minSup_u ) → addTo(INF)
```

In the last case, items in INF are discarded because they won't be consider like an useful information but perhaps for another incremental mining.

In the case of new coming uDB, the bounder of promising items will be calculate using the last size of DB ($|nDB|$).

Health Check process

In spite of keeping promising items and minimize discarding others, accumulation of infrequent items (INF set) can germinate and probably contain frequent itemset and/or promising items. So, at which moment should we revise accumulated infrequent items? Detect this point of deviation can be get using Machine Learning Techniques, in our case we propose to use Artificial Neural Network (*ANN*) [20]. In learning phase, we have to feed NN with size of frequent *1*-itemset obtained by IncA algorithm, size of DB and size of infrequent items INF; as output we given correct or incorrect situation according to frequent *1*-itemset obtained by classic Apriori. After executing learning phase, NN will be able to detect deviation point using only obtained frequent *1*-itemset, DB and accumulated infrequent items (Fig. 2).

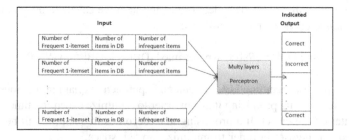

Fig. 2. ANN – learning phase

We call the process of checking accumulated infrequent items a Health Check (HC). Incremental Clustering Apriori hurl deviation check (send suitable input to NN) after each update, if output is 'correct' we keep obtained frequent itemset and promising items; if output is 'incorrect' a HC is lance to update obtained frequent itemset and promising items. The aim of health check process is to assure that all useful items are considered even those already discarded, all ensuring a minimum executing time and optimal resource consummation.

4 Experiments and Results

In order to check IncA performance, we implemented the proposed algorithm in Java and we used MATLAB to simulate the *HC* process. To make experimentation we used two synthetic dataset CACM and RCV1:

CACM database contain a set of articles, it comprises 3204 transactions and around 2500 different items. RCV1 database is a text categorization test collection, it comprises 18758 transactions and around 500000 different items (Fig. 3).

For each test we have taken 20% of database as insertion and we have varied support. KPI: number of generated frequent itemset and execution time (Tables 1 and 2).

Table 1. Analyses by number of frequent itemsets and promising itemsets

	minSup = 60%				minSup = 40%				minSup = 20%			
	Apriori		IncA		Apriori		IncA		Apriori		IncA	
	|F|	|PF|	|F|	|PF|	|F|	|PF|	|F|	|PF|	|F|	|PF|	|F|	|PF|
CACM	11	0	11	36	48	0	48	87	93	0	93	169
RCV1	44	0	44	27	85	0	83	50	133	0	132	205

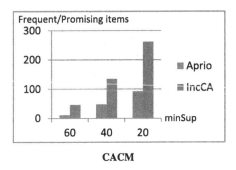

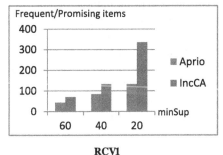

CACM **RCV1**

Fig. 3. Frequent/Promising items number applying IncA

Table 2. Analyses by execution time evolution

	minSup = 60%		minSup = 40%		minSup = 20%	
	Apriori	IncA	Apriori	IncA	Apriori	IncA
	Time (s)		Time (s)		Time (s)	
CACM	39600	14400	50400	21600	97200	75600
RCV1	64800	36000	82800	61200	129600	118800

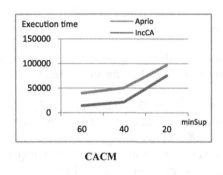

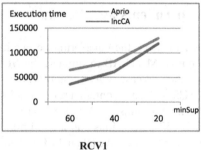

CACM   RCV1

Fig. 4. Execution time applying IncA

Performed experiments have showed that IncA store more items than classic Apriori, but considered frequent item number is always a same as classic Apriori. Furthermore, observing running time (Fig. 4) IncA performs clearly more than classic Apriori either when changing support.

Otherwise, lose frequent items is not possible and running time is minimized by applying IncA.

Likewise, if we compare IncA with other techniques, we remark a better performances: according to FUP experimentations [13] this algorithm is faster than Apriori but there is a reduction in the number of candidate, but IncA is also faster because it pure 1-itemsets and skip purning phase for 2-itemsets. EDUA [14] is use scan reduction technique, it kept an optimal number of 2-itemset -as mentioned in experiment part- and generate a less number of frequent itemsets than IncA which kept frequent and promising itemsets. Comparing IncA with a probability approach [15], this latter uses a bounder based on probabilities and maximize number of kept candidate items as we can see in experimentations, so memory requirement is more considerable than the memory space occupied by IncA. Notice also that IncA is adaptive to any size of incremental database unlike a probability approach that required knowing incremental data size in order to define bounders which is not convenient.

5 Conclusion

In this paper, we proposed a new version of Apriori algorithm for evolving databases called Incremental Apriori (*IncA*). It consists in scanning incoming transactions and updating itemsets frequencies without discarding any potential information by keeping promising items and applying a health check process. Unlike the existing efforts, the proposed approach operates efficiently in large dataset insertion and an ameliorated version can be efficient for deletion case. IncA develops several advantages as saving running time, saves the costly database and eliminate information loss. Performed experimentations on synthetic data show that the developed algorithm is largely more efficient than the classic Apriori and than probability based IARD.

The proposed approach presents a novel incremental approach to adopt to more applications in mining data. As perspectives, our approach can be applied for data reduction and databases representation. On the other hand, we plan to adopt IncA on insertion/deletion case simultaneously and test it on real data. We are also thinking about enriching IncA with a technique to reduce candidate itemsets size and enhance the scalability of this approach regarding to association rules generation and improvement.

References

1. Jiewai, H., Kamber, M.: Data Mining: Concepts and Techniques. Morgann Kaufmann, San Francisco (2011)
2. Information overload. Nature **460**, 551 (2009). doi:10.1038/460551a. Accessed 29 July 2009
3. Leung, C.K.-S., Khan, Q.I., Li, Z., Hoque, T.: CanTree: a canonical-order tree for incremental frequent pattern mining. Knowl. Inf. Syst. **11**(3), 287–311 (2007)
4. Rakesh, A., Ramakrishnan, S.: Fast algorithms for mining association rules. In: 20th International Conference on Very Large Data Bases, Chile, pp. 487–499 (1994)
5. Bastide, Y., Taouil, R., Pasquier, N., Stumme, G., Lakhal, L.: Pascal: un algorithme d'extraction des motifs fréquents, pp. 65–95. Techniques et Sciences Informatiques, Editions Hermès (2002)
6. Han, J.L., Plank, A.W.: Background for association rules and cost estimate of selected mining algorithms. In: 5th International CIKM, USA, pp. 73–80 (1996)
7. Zhang, S., Zhang, J., Zhang, C.: EDUA an efficient algorithm for dynamic database mining. Inf. Sci. **177**, 2756–2767 (2007)
8. Jiemin, Z., Defu, Z., Leung, S.C.H., Xiyue, Z.: An efficient algorithm for frequent itemsets in data mining. In: ICSSSM, Hong Kong, pp. 1–6. IEEE (2010)
9. Khan, Z., Faujdar, N., Singh, P., Abbas, T.: Modified Bitapriori algorithm: an intelligent approach for mining frequent Ite-Set. In: International Conference on Advance in Signal Processing and Communication, India, pp. 813–819 (2013)
10. Park, J.S., Chen, M.S., Yu, P.S.: An effective hash-based algorithm for mining association rules. In: Proceedings of 1995 ACM SIGMOD International Conference on Management of Dai, San Jose, pp. 175–186 (1995)
11. Cheung, D.W., Han, J., Ng, V.T., Wong, C.Y.: Maintenance of discovered association rules in large database: an incremental updating technique. In: Proceedings of 12th IEEE International Conference on Data Engineering, pp. 106–114 (1996)
12. Suresh, P., Nithya, K.N., Murugan, K.: Improved generation of frequent item sets using apriori algorithm. IJARCCE Int. J. Adv. Res. Comput. Commun. Eng. **4**(10) (2015)
13. Cheung, D.W., Lee, S.D., Kao, B.: A general incremental technique for mining discovered association rules. In: Proceedings of International Conference on Database System for Advanced Applications, pp. 185–194 (1997)

14. Lee, C., Lin, C.R., Chen, M.S.: Sliding-window filtering: an efficient algorithm for incremental mining. In: Proceedings of International Conference on Information and Knowledge Management, CIKM01, pp. 263–270 (2001)
15. Thusaranon, P., Kreesuradej, W.: A probability based incremental association rule discovery. In: 19th International Symposium on Artificial Life and Robotics, Oita, pp. 22–24. Department of Information System, Information Technology Faculty *Dhurakij* Pundit University, Thailand (2014)
16. Yao, Y.Y.: Three-way decision with probabilistic rough sets. Inf. Sci. **180**, 341–353. Department of Computer Science, University of Regina, Regina, Saskatchewan, Canada (2010)
17. Wen, P., Li, Y., Polkowski, L., Yao, Y.Y., Tsumoto, S. (eds.): Rough Sets and Knowledge Technology: 4th International Conference, RSKT 2009. LNCS, vol. 5589, pp. 642–649. Springer, Heidelberg (2009)
18. Yao, Y.Y: Decision-theoretic rough set models. In: Yao, J., Lingras, P., Wu, W.-Z., Szczuka, M.S., Cercone, N.J., Ślęzak, D. (eds.) RSKT 2007. LNCS (LNAI), vol. 4481, pp. 1–12. Springer, Heidelberg (2007)
19. Dong, J., Han, M.: BitTableFI an efficient mining frequent itemsets algorithm. Knowl. Based Syst. **20**(4), 329–335 (2007)
20. Niknafs, A., Parsa, S.: A neural network approach for updating ranked association rules, based on data envelopment analysis. J. Artif. Intell. **4**, 279–287 (2011). Department of Computer Engineering, Shahid Bahonar University of Kerman, Iran, Asian Network for Scientific Information, Iran
21. Hegland, M.: The apriori algorithm – a tutorial. In: Mathematics and Computation in Imaging Science and Information Processing, vol. 11, pp. 209–262. World Scientific Publishing (2007)

Multi-objective Bat Algorithm for Mining Interesting Association Rules

Kamel Eddine Heraguemi[1(✉)], Nadjet Kamel[1], and Habiba Drias[2]

[1] Depart. Computer Science, Fac-Sciences, Univ-Setif, Setif, Algeria
k.heragmi@univ-setif.dz, nkamel@usthb.dz
[2] Depart. Computer Science, USTHB, Algiers, Algeria
hdrias@usthb.dz

Abstract. Association rule mining problem attracts the attention of researchers inasmuch to its importance and applications in our world with the fast growth of the stored data. Association rule mining process is computationally very expensive because rules number grows exponentially as items number in the database increases. However, Association rule mining is more complex when we introduce the quality criteria and usefulness to the user. This paper deals with association rule mining issue in which we propose Multi-Objective Bat algorithm for association rules mining Known as MOB-ARM. With the aim of extract more useful and understandable rules. We introduce four quality measures of association rules: Support, Confidence, Comprehensibility, and Interestingness in two objective functions considered for maximization. A series of experiments are carried out on several well-known benchmarks in association rule mining field and the performance of our proposal are evaluated and compared with those of other recently published methods including mono-objective and multi-objective approaches. The outcomes show a clear superiority of our proposal in-face-of mono objective methods in terms generated rules number and rule quality. Also, The analysis also shows a competitive outcomes in terms of quality against multi-objective optimization methods.

Keywords: Association rules mining · ARM · Bat algorithm · Multi-objective optimization · Support · Confidence · Comprehensibility · Interestingness

1 Introduction

Association rule mining [1] (ARM) is one of the most active, attractive and useful research area in knowledge discovery. Basically, it finds practical and interesting relations between items in huge transactional databases to help for decision making. The extracted relationships can be represented by *IF-THEN* statement, *IF <some conditions are satisfied> THEN <some values of other attributes>*. The conditions in *IF* statement called *Antecedent* and those within the *THEN* clause are *Consequence*. ARM applications varies from market basket analysis

© Springer International Publishing AG 2017
R. Prasath and A. Gelbukh (Eds.): MIKE 2016, LNAI 10089, pp. 13–23, 2017.
DOI: 10.1007/978-3-319-58130-9_2

as first innovation [1] toward more important and hypersensitive fields, such as: business intelligence, Medical and natural language processing, which make ARM process and relationships among attributes of datasets indispensable.

Nowadays, discovery such relationships in large database is NP-Complete problem [3]. The huge quantity of stored data makes classical approaches applied to extract association rules in such database running slowly. In these methods the growth of features number results in a dramatic increase of the processing time. This is why researchers in ARM headed to optimization with intelligent algorithms which presents robust and efficient approaches to explore a massive search space. Intelligent algorithms have already shown their efficiency to solve combinatorial problems (NP-Complete). Generally, an evolutionary algorithm maintains a population of individuals; each one represents a solution to the given problem. Each individual is evaluated by a fitness function which determines solution quality. Assuming that databases are simple search spaces, same concepts exist in association rule problem where the algorithm maintains a set of rules which are individuals and evaluate them using different quality measures (confidence, comprehensibility, interesting ...). The most popular intelligent methods that have been applied to ARM problem are: Genetic algorithm, particle swarm algorithm, bees swarm algorithm, and bat algorithm. Most of these approaches deal with the ARM problem as a single-objective optimization problem. However, they still generate useless rules for decision making process because they utilize just support and confidence to evaluate the rules. Recently, many works on association rules deal with ARM as multi-objective optimization perspective to extract a small set of useful and comprehensible rules by introducing several measures in assessment of rules.

In this paper we propose a multi-objective method to mine interesting and useful association rules within transactional databases, starting from a minimum support and confidence threshold specified by the final users according to their needs, based on the multi-objective bat algorithm. In order to improve the efficiency of our algorithm some new contributions have been embedded in our proposal. We use four measures to evaluate extracted rules quality and define two global objective functions considered for optimization (maximization) in order to extract better promising association rules.

The rest of this paper is organized as follows. The next section presents a general background on multi-Objective Optimization and association rule mining problems. Also, we recall the modified bat algorithm for association rule mining (BAT-ARM). Section 3 leads with summary of the existing ARM algorithms. Section 4, presents formally our approach. Section 5, reports on the experimental results for our approach and the comparison with other ARM existing algorithms. Finally, we conclude with our prospective for a future work.

2 Preliminaries

2.1 Multi-objective Optimization Problems

A general Multi-Objective Optimization problems (MOOP) includes a set of n parameters(decision variables), a set of k objective functions, and a set of m

constraints. Objectives and constraints are functions of the decision variables. Generally, The form of a MOOP can be described as minimizing/maximizing a set of objective functions, $f(x) = (f_1(x), ..., f_k(x))^T$ subject to $G_i(x) <= 0, i = 1, 2, ..., n$, by finding the vector $X = (x_1, x_2, ..., x_n)^T$. It is noted that $G_i(x)$ are the constraints that must be satisfied while minimizing/maximizing the objective functions. With the presence of several objective functions, the notion of "optimum" has changed to "Pareto optimum" [5] because MOOP aims to find a vector of solutions rather than a single solution. In general, it is not possible to find an exact PF for complex MOOPs, and in such cases the goal is to determine a Pareto optimal set that approximates the exact PF as close as possible by generating a diverse range of solutions.

2.2   Association Rule Mining

Formally, association rule [1] problem is defined as follow: Let $I = \{i_1, i_2, ..., i_n\}$ be a set of literals called items, let D be a transactional database where each transaction T contains a set of items. An association rule is implication like $X \implies Y$ where $X, Y \in I$ and $X \cap Y = \emptyset$. The item-sets X, Y are named antecedent and consequent, respectively. Mainly, two principal measures are used to detect the interesting and useful association rules: Support and Confidence. They are defined as follows:

Support: written supp(X), it is the proportion of transactions in D that contains X, to the total of records in database. Support is calculated using the following equation

$$supp(X) = |\{(y, Xy) \in D/X \subseteq Xy\}|/|D| \tag{1}$$

The support of an association rule $X \to Y$ is the support of $X \cup Y$.

Confidence: written conf($X \to Y$), it is the proportion of transactions covering X and Y, to the total of records containing X, when the percentage exceeds threshold of confidence an interesting association rule can be generated. The confidence of a rule is calculated as:

$$conf(X \to Y) = supp(X \cup Y)/supp(X) \tag{2}$$

In another word, support denotes the frequency of occurring patterns while confidence expresses the strength of implication [1].

2.3   ARM Based on Bat Algorithm (BAT-ARM)

In [11], we proposed a new algorithm for ARM inspired from bat behavior, which aims to generate the best rules in defined dataset starting from minimum support and confidence with reasonable execution time. In association rule mining, the rule is accepted if its support and confidence satisfy user minimum support and confidence threshold. Based on this definition we describe a simple objective function based on the support and the confidence to evaluate the solution and never generate invalid rules. Based on the definition of bat algorithm in [18] new formal description for bat motion is described related to association rule mining bases, where the frequency, velocity and position are defined as follow:

- **Frequency** f_i: presents how many items can be changed in the actual rule, where the maximum frequency f_{max} is the number of attributes in the dataset and the minimum frequency f_{min} is 0.
- **Velocity** v_i: indicates where the changes will be started.
- **Position** x_i: it is the new generated rule based on new frequency, velocity and the loudness.

The generation of new positions (rules) is extracted based on the frequency, velocity of each virtual bat which are updated at each iteration by Eqs. 3 and 4

$$f_i^t = 1 + (f_{max})\beta, \tag{3}$$

$$v_i^t = f_{max} - f_i^t - v_i^{t-1}, \tag{4}$$

Generally, BAT-ARM provides a great performance in term of CPU-time and memory usage in the face of FP-growth algorithm, thanks to the echolocation concept of the bat algorithm that can determine which part of the best rule have changed to get a better position (rule) for the actual bat. However, the ARM by means of single objective optimization methods also has a few limits that are listed as follows. Firstly, in order to solve association rule mining issues, these methods explore the maximum of search space, which generates many rules having a high fitness value. This process can generate many redundant rules. Secondly, as this method focuses on covering the search space and generating the maximum of rules, it neglects the comprehensibility and usability of rules that are meaningful for the end users. To overcome these main drawbacks we investigate with a new approach based on multi-objective bat algorithm.

3 Literature Review

This section presents a literature review on the existing evolutionary algorithms that deal with ARM issue. In the literature, there are many other bio-inspired approaches are proposed to extract association rules, In [17] G3PARM algorithm is developed, it is based on genetic programming. The authors used grammar guided genetic programming (G3P) to avoid invalid individuals found by Genetic Programming (GP) process. Also, G3PARM permits multiple variants of data by using a context free grammar. In [6] the authors developed a new approach inspired from bees behavior and based on bee swarm optimization algorithm called BSO-ARM. The results of this approach show that BSO-ARM performs better than all genetic algorithms. As extension to their work, the authors present an amelioration to BSO-ARM in [7], where three strategies to determine the search area of each bee are proposed (modulo, next, syntactic). In our earlier work [11] we present an adaptation of bat algorithm to association rule mining issue known as BAT-ARM. We present a new mathematical definition for the virtual bats motion related to ARM problem basics. The outcomes show a high performance in solution quality and CPU-time consumption thanks to the echolocation concept of bat algorithm. Later, within [12] we propose a multi-population bat algorithm to extract association rules within transactional database which is based on the search process developed in BAT-ARM. Furthermore, sub-populations use master-slave plan to cooperate among themselves. The results outperform those of BAT-ARM in both quality and time execution. This later was improved by new cooperation strategies in [13]. All these motioned methods are single objective approaches which stay suffer from several drawbacks mainly the huge number of generated rules and the extraction of useless ones.

These disadvantages open the door to dealt with association rule mining as a multi-objective optimization problem where different measures are used in the same algorithm. In [15], a multi-objective genetic algorithm approach to mine association rules for numerical data was proposed, where confidence, interestingness and comprehensibility are used to define the fitness function. Results showed that the generated rules are more appropriate than similar approaches. In [16] the authors proposed a multi-objective genetic algorithm for generating interesting association rules with multiple criteria i.e. support, confidence and simplicity (comprehensibility). Their method can identify the interesting rules without having the user-specified thresholds of minimum support and minimum confidence. Another study presented in [4] discussed multi-objective particle swarm optimization algorithm for numerical ARM named MOPAR. This method uses confidence, comprehensibility, and interestingness to evaluate the extracted rules. In [8], three multi-objective techniques proposed for mining association rules without specifying neither support nor confidence by optimizing several quality measures. The methods are Multi-objective Binary Particle Swarm Optimization (MO-BPSO), a Multi-objective Binary Firefly optimization and Threshold Accepting (MO-BFFO-TA), and a Multi-objective Binary Particle Swarm optimization and Threshold Accepting (MO-BPSO-TA). More recently, a new multi-objective evolutionary algorithm, MBAREA, for mining useful Boolean association rules with low computational cost is proposed in [14].

4   Multi-objective Bat Algorithm for ARM

4.1   Rule Encoding

In our method we use Michigan Approaches. Where each solution X represents a rule that contains k items. Therefore, the solution X is represented with a vector S which contains $k + 1$ positions where:

1. S [0] separates between the antecedent and the consequent of the rule,
2. S[i] = j where i > 0 If the jth item in the database is in the rule, else the position contains 0.

For example, let $I = \{i_1, i_2, ..., i_{10}\}$ be a set of items:

- $X1 = \{3, 1, 5, 0, 6, 2, 0, 0, 7, 0, 0\}$ represents the rule $i_1, i_5 \Rightarrow i_6, i_2, i_7$,

4.2   Objective Functions

As mentioned above single objective evolutionary algorithms use generally only one measure i.e., *Support, Confidence, etc.* to evaluate extracted rules quality. These measures assess rules depending on number of occurrence in database. Nevertheless, these algorithms do not give any importance to other rule quality measures like i.e., comprehensibility and interestingness.

In our work we use *comprehensibility, interestingness* measures in addition to confidence and support which are used as fitness function in [11], and as objectives in our method (MOB-ARM). The confidence criterion evaluates the quality of each rule based on occurrences number in the whole dataset. When the rule has more occurrences in the database, this means the rule has a better quality. We define the first objective for our method using the support and confidence, shown in Eq. 5.

$$Obj_1(R) = \alpha conf(R) + \beta supp(R)/\alpha + \beta \tag{5}$$

When the rule contains a huge number of attributes, this makes the rule more difficult to comprehend. If generated rules are not comprehensible for the user, they will be useless. This is why we introduce the comprehensibility measure, which can be modeled as shown in Eq. 6.

$$Comprehensibility(X) = \frac{\log(1+|Conseq|)}{\log(1+|Antec \cup Conseq|)} \tag{6}$$

Where, $|Conseq|$ and $|Antec \cup Conseq|$ are items number in the Consequence part and the total rule respectively. The comprehensibility increases and the rules are more understandable whenever items number in the antecedent part was smaller. Moreover, interestingness of a rule is used to quantify how much rule is surprising for users. As the most important point of rule mining is to find some hidden information, it should discover those rules having comparatively less occurrence in the database. Interestingness measure is defined by Eq. 7.

$$Interesting(X) = \frac{Supp(A \cup C)}{Supp(A)} \times \frac{Supp(A \cup C)}{Supp(C)} \times \frac{(1 - Supp(A \cup C))}{N} \tag{7}$$

Where A, C and N are the antecedence, consequence and transactions number in the whole database, respectively. We define the second objective for our algorithm based on *Comprehensibility, Interestingness* using Eq. 8.

$$Obj_2(R) = \gamma Comp(R) + \delta Inter(R)/\gamma + \delta \tag{8}$$

Where α, β, γ and δ are empirical parameters which are chosen relative to the importance of support, confidence, Comprehensibility and Interestingness to final user.

Algorithm 1. MOB-ARM Algorithm pseudo code

---

objective functions $f_1(x), ..., f_k(x)$.
Initialize the bat population x_i and v_i;
Initialize pulse rates r_i and the loudness A_i;
for $j = 1$ to N (points on Pareto fronts) **do**
  Generate K weights $w_k \geq 0$ so that $\sum_{k=1}^{k} w_k = 1$;
  Form a single objective $\sum_{k=1}^{k} w_k f_k$;
  while ($t < $ *Max number of iterations*) **do**
    Generate new solutions by adjusting frequency f_i;
    and updating velocities and locations/solutions [11];
    Generate a new solution x_i [11];
    if ($rand > r_i$) **then**
      Generate a local solution around the selected best solution by changing only one item in the rule;
    end if
    if ($f(x_i) > f(x_i*)$) **then**
      Accept the new solutions;
      $x_i* = x_i$;
      Increase r_i and reduce A_i
    end if
    Rank the bats according to the best solution;
  end while
  Record x_i* as non-dominate solution;
end for
Post-process results and visualize the best detected rules.

4.3   The Algorithm Flow

Algorithm 1 illustrates the pseudo code of MOB-ARM. The main computational steps of the proposal are described as follows:

- **Initialization step:** firstly, all the bats are initialized with random frequency and velocity. The values are taken in the intervals [0, items number] and [0, items number+1] respectively. A randomly generated position/solution (rule) is affected to each bat, and an initial rate and loudness is affected to each bat randomly.
- **Search the non-dominate solution for the Pareto point:** For each Pareto point, a new single global objective function is generated based on weights w_k in which their sum is equal to 1 ($\sum_{k=1}^{k} w_k = 1$). The global objective function is generally presented by: $\sum_{k=1}^{k} w_k f_k$, where k is objective functions number used for the mining problem. In our case, we have only two objective functions presented in Eqs. 5 and 8. So the global function is defined as follows:

$$Obj(R) = w_1.Obj_1(R) + w_2.Obj_2(R); \qquad (9)$$

- **Search the best solution (Rule) for each bat at the Pareto point:** At each iteration, a new rule is generated based on BAT-ARM described in [11] by adjusting frequency, updating velocity and location. If the new fitness is better than the previous one, then the rule will be accepted, the loudness A_i reduced, and the rate r_i increased according to updating equations in [18].

5   Experimental Results

In order to perform experimentations, several well-known and frequently used real world datasets in data mining, such as Frequent and mining dataset Repository [9], Bilkent University Function Approximation Repository [10], are used in this section for several tests. This section describes the used benchmarks. After-that, comparative study with BAT-ARM and MPB-ARM, which are two mono-objective versions of bat algorithm updated to association rules mining, is given. Also, we present comparison of our approach to three other multi-objective methods recently published. All algorithms are written in Java and executed on Intel core I5 machine with 4 GB of memory running under Linux Ubuntu. We examined our approach on seven well known datasets with different sizes of transactions, items and average size per transaction. For instance, Chess dataset has 3196 transactions with 75 items when the average per transaction is 37, unlike mushroom dataset which has much more transactions and items when it has just 23 items per transaction. Table 1 presents different datasets used in our experiments.

Table 1. Description of experimental benchmark

Dataset	Transactions size	Item size
Basketball	96	5
Bodyfat	252	15
Quake	2178	4
IBM Quest Standard	1000	20
Chess	3196	37
Mushroom	8124	23

5.1   Comparative Study to Single Objective Approaches

In this section, we propose a study that compares our new approach to single objective versions of bat algorithm designed for mining association rules (BAT-ARM, MPB-ARM). This experiment was cried on three datasets with medium transactions size (*IBM Quest Standard, Chess and Mushroom*). The default parameters of the BAT-ARM, MPB-ARM and MOB-ARM are defined to make the comparison completely fair where support and confidence thresholds are fixed to 0.2 and 0.5 respectively.

Table 2 presents the average results of thirty executions on three algorithms (BAT-ARM, MPB-ARM and MOB-ARM). In our comparison, three axes are

Table 2. Comparison of results to mono-objective methods in terms of number of generated rules, support and confidence

	Algorithms	Datasets		
		IBM-standard	Chess	Mashroom
No. of rules	MOB-ARM	215	293	26
	BAT-ARM	485	1870	341
	MPB-ARM	850	739	791
Support (%)	MOB-ARM	26	51	34
	BAT-ARM	25	38	23
	MPB-ARM	23	46	23
Confidance (%)	MOB-ARM	54	83	87
	BAT-ARM	52	72	54
	MPB-ARM	59	79	78.5

taken into account: average support, confidence and number of generated rules. Outcomes shows that the proposed algorithms extract less number of rules for all the datasets. This is because new criteria of selection are introduced as objectives (Comprehensibility and interestingness), so MOB-ARM generates only useful and understandable rules for the user. On contrary, mono-objective approaches generate the maximum number of rules that satisfy support and confidence thresholds. In terms of support and confidence we note that MOB-ARM is more robust than BAT-ARM and MPB-ARM because of the small number of extracted rules and dominance conditions applied when mining association rules.

5.2 Comparative Study to Multi-objective Approaches

In this study, effectiveness of the proposed algorithm is compared with three similar algorithms. All of these methods are based on a multi-objective evolutionary approach and designed for association rule mining. The three algorithms are: MODENAR [2], MOGAR [15] and MOPAR [4].

Table 3 compares the outcomes obtained by MOB-ARM to previous similar methods that deal with association rule mining as multi-objective optimization problem in terms of average support. The results show that our proposed method yields a competitive support of the extracted rule. However, in case of Bodyfat datasets the overage support is less than other methods (MODENAR, MOGAR), this is caused by the fact of the strict application of dominance conditions.

In addition, we compute the average confidence to evaluate the strength of extracted rules. Table 3 shows that our suggested method gives acceptable results. Our results can be improved and give better confidence average because we use a minimum confidence threshold that can be changed by the user according to his exigencies. To make the study more comprehensive, we calculate the average number of extracted rules for each dataset and the results are presented in Table 3. From the outcomes, we observe that our method have a stable behavior and it is competitive to the previous methods.

Table 3. Comparison of results to mono-objective methods in terms of number of generated rules, support and confidence

	Algorithms	Datasets		
		Basketball	Bodyfat	Quake
Support (%)	MOPAR	30.76	22.95	31.97
	MODENAR	37.20	65.22	39.86
	MOGAR	50.85	57.22	30.12
	MOB-ARM	37.5	23	41
Confidance (%)	MOPAR	95	81.8	89.32
	MODENAR	61	62	63
	MOGAR	83	85	79
	MOB-ARM	79	83	88
No. of rules	MOPAR	69	70	54
	MODENAR	48	52	55
	MOGAR	50	84	44
	MOB-ARM	63	51	50

6   Conclusion

In this paper, we proposed a new multi-objective meta-heuristic to deal with association rule mining based on bat algorithm and called MOB-ARM. The proposal uses four quality measures, (Support, Confidence, Comprehensibility and Interestingness) to extract the best rules that help the user in decision making process and can be understood. Our approach is based on vertical dataset representation that reduces the computation time of computing and avoids the repeated scans of the whole datasets in each rule evaluation. The performance of MOB-ARM has been compared to two single objective algorithms based bat algorithm for mining association rules and three other methods dealing with multi-objective miners. The experimental results prove the effectiveness of our proposed method. For the near future, we aim to develop a new version that deals with quantitative association rules without requiring a discretization step. We think also about parallelism the algorithm and implement it on a GPU to improve both solution quality and running time.

References

1. Agrawal, R., Imieliński, T., Swami, A.: Mining association rules between sets of items in large databases. In: ACM SIGMOD Record, vol. 22, pp. 207–216. ACM (1993)
2. Alatas, B., Akin, E., Karci, A.: Modenar: multi-objective differential evolution algorithm for mining numeric association rules. Appl. Soft Comput. 8(1), 646–656 (2008)

3. Angiulli, F., Ianni, G., Palopoli, L.: On the complexity of mining association rules. In: SEBD, pp. 177–184 (2001)
4. Beiranvand, V., Mobasher-Kashani, M., Bakar, A.A.: Multi-objective PSO algorithm for mining numerical association rules without a priori discretization. Expert Syst. Appl. **41**(9), 4259–4273 (2014)
5. Coello, C.A., Coello, D.A., Veldhuizen, V., Lamont, G.B.: Evolutionary Algorithms for Solving Multi-Objective Problems, vol. 242. Springer, Heidelberg (2002)
6. Djenouri, Y., Drias, H., Habbas, Z., Mosteghanemi, H.: Bees swarm optimization for web association rule mining. In: 2012 IEEE/WIC/ACM International Conferences on Web Intelligence and Intelligent Agent Technology (WI-IAT), vol. 3, pp. 142–146. IEEE (2012)
7. Djenouri, Y., Drias, H., Habbas, Z.: Bees swarm optimisation using multiple strategies for association rule mining. Int. J. Bio-Inspired Comput. **6**(4), 239–249 (2014)
8. Ganghishetti, P., Vadlamani, R.: Association rule mining via evolutionary multi-objective optimization. In: Murty, M.N., He, X., Chillarige, R.R., Weng, P. (eds.) MIWAI 2014. LNCS (LNAI), vol. 8875, pp. 35–46. Springer, Cham (2014). doi:10.1007/978-3-319-13365-2_4
9. Goethls, B., Zaki, M.J.: Frequent itemset mining dataset repository (2003). http://fimi.ua.ac.be/data/
10. Guvenir, H.A., Uysal, I.: Bilkent university function approximation repository (2000). http://funapp.cs.bilkent.edu.tr/DataSets/
11. Heraguemi, K.E., Kamel, N., Drias, H.: Association rule mining based on bat algorithm. J. Comput. Theor. Nanosci. **12**(7), 1195–1200 (2015)
12. Heraguemi, K.E., Kamel, N., Drias, H.: Multi-population cooperative bat algorithm for association rule mining. In: Núñez, M., Nguyen, N.T., Camacho, D., Trawiński, B. (eds.) ICCCI 2015. LNCS (LNAI), vol. 9329, pp. 265–274. Springer, Cham (2015). doi:10.1007/978-3-319-24069-5_25
13. Heraguemi, K.E., Kamel, N., Drias, H.: Multi-swarm bat algorithm for association rule mining using multiple cooperative strategies. Appl. Intell. **45**(4), 1021–1033 (2016). Springer
14. Kabir, M.M.J., Xu, S., Kang, B.H., Zhao, Z.: A new evolutionary algorithm for extracting a reduced set of interesting association rules. In: Arik, S., Huang, T., Lai, W.K., Liu, Q. (eds.) ICONIP 2015. LNCS, vol. 9490, pp. 133–142. Springer, Cham (2015). doi:10.1007/978-3-319-26535-3_16
15. Minaei-Bidgoli, B., Barmaki, R., Nasiri, M.: Mining numerical association rules via multi-objective genetic algorithms. Inf. Sci. **233**, 15–24 (2013)
16. Al-maqaleh, M.B.: Discovering interesting association rules: a multi-objective genetic algorithm approach. Int. J. Appl. Inf. Syst. **5**(3), 47–52 (2013)
17. Olmo, J.L., Luna, J.M., Romero, J.R., Ventura, S.: Association rule mining using a multi-objective grammar-based ant programming algorithm. In: 2011 11th International Conference on Intelligent Systems Design and Applications (ISDA), pp. 971–977, November (2011)
18. Yang, X.-S.: A new metaheuristic bat-inspired algorithm. In: González, J.R., Pelta, D.A., Cruz, C., Terrazas, G., Krasnogor, N. (eds.) Nature Inspired Cooperative Strategies for Optimization (NICSO 2010). Studies in Computational Intelligence, vol. 284, pp. 65–74. Springer, Heidelberg (2010). doi:10.1007/978-3-642-12538-6_6

Dynamics of Self-replicating DNA-Tile Patterns

Vinay Kumar Gautam[✉] and Rajendra Prasath

Department of Computer and Information Science, The Norwegian
University of Science and Technology, Sem Sælands Vei 7–9, 7491 Trondheim, Norway
vkgautam@idi.ntnu.no, drrprasath@gmail.com

Abstract. DNA tiles serve as molecular components for the self-assembly of programmable 2-dimensional patterns at the nanoscale. To produce identical copies of a pre-assembled DNA tile pattern, we use a theoretical framework of non enzymatic cross-coupled self-replication system based on tile self-assembly model. This paper presents a kinetic modelling of the pattern self-replication and analyses the influence of physicochemical parameters of tile self-assembly process over the reliability and replication gain. We demonstrate that the tile assembly errors, introduced in tile patterns during their assembly, set a limit over the size of a tile pattern that can be replicated exponentially and reliably.

Keywords: Self-replication · DNA tile · DNA self-assembly · Tile pattern

1 Introduction

Self-replication is a fundamental mechanism in biology, that nature applies to create elegant molecular systems inexpensively using processes of growth and selection. Application of this inspiration to engineer artificial molecular systems has been a constant pursuit of nanosciences. Gunter von Kiedorowski [10] first introduced a minimal system of molecular self-replication, which typically involves a three-step process. First, a template molecule assembles with few substrate molecules resulting in an intermediate complex formation. Second, the substrate molecules within the complex join together irreversibly by covalent binding, and thereby forming a replica of the template. Third, the complex molecule dissociates into two templates: the former template and the newly created replica. Each of these templates can reiterate the three-step process adding to the template population. The population have been observed to grow sub-exponentially (parabolic) [16].

Template directed non-enzymatic self-replication has been used for the synthesis of nucleic acid sequences using only linear organization of short sequences of nucleic acids (primers). However, recent advances in structural DNA self-assembly have opened up perspectives for the non-enzymatic self-replication of two-dimensional (2-D) and three-dimensional (3-D) patterns of

R. Prasath—A part of this was carried out when the author was in Indian Institute of Information Technology (IIIT) Sricity, India.

R. Prasath and A. Gelbukh (Eds.): MIKE 2016, LNAI 10089, pp. 24–37, 2017.
DOI: 10.1007/978-3-319-58130-9_3

DNA [1,9,14,18,23]. A 2-D DNA pattern replication based on crystal growth followed by random splitting has been experimentally demonstrated for the amplification of combinatorial information [15].

DNA tile self-assembly [20] is an emerging paradigm for nanostructure construction and molecular scale computation. DNA tiles [19], the building blocks of tile self-assembly, can be designed to interact with strength and specificity for the assembly of logically and/or algorithmically directed periodic and aperiodic 2-D intricate patterns. For a theoretical modelling of tile assembly, Erik Winfree first introduced an abstract Tile Assembly Model (aTAM) [20]. In the aTAM framework, the assembly starts from a single seed tile and the pattern grows in 2-D as more tiles adjoin one-by-one following a simple assembly rule — the total binding strength of an incumbent tile should be greater than or equal to a threshold value known as temperature parameter of assembly. However, DNA tile assembly is essentially a physico-chemical process, where local reaction temperature and tile concentration are the governing factors. Therefore, for a realistic modelling of tile assembly process, Winfree introduced kinetic Tile Assembly Model (kTAM) [21]. The kTAM considers each tile assembly step as a reversible process governed by the tile concentration, local reaction temperature and binding strengths of tiles. The model enables analysis of the assembly errors and growth rate for a given tile assembly system.

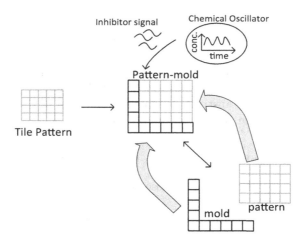

Fig. 1. A simplistic view of the tile pattern self-replication system: the self-replication starts with a pre-assembled target tile pattern (a 4×5 tile pattern is shown in gray); a mold (shown in black) assembles around the South-West border of the pattern; a cyclically generated inhibitor signal dissociates the mold and pattern, which initiate a cross-coupled cycle of pattern self-replication.

A very high level abstract version of tile pattern self-replication system [6], shown in Fig. 1, is designed using additional tiles, which self-assemble to form a mold structure around the L-shaped South-West border of the target pattern. The assembled mold consists of switching enabled tiles that are dynamically triggered by an externally supplied inhibitor signal of DNA to dissociate the pattern and mold templates. The dissociated mold and pattern structures further catalyse the assembly of new templates of patterns and mold structures, respectively. The inhibitor signal is cyclically released by a chemical oscillator tuned to the time intervals involved in the mold formation and pattern formation. Thus, the entire process forms the basis of a cross-coupled self-replication system of 2-D patterns of tiles.

In this paper, we derive a kinetic model for the population growth of pattern self-replicator using chemical kinetic rates of tile assembly and disassembly in the kTAM. Kinetic rates of tiles are governed by physicochemical parameters (local assembly temperature, total binding strength and concentration of tiles), which causes erroneous assembly of tiles. We analyse the impact of these parameters in population growth dynamics of the pattern self-replicator and reliability of the replicated patterns. Population growth, fidelity and size of replicating patterns are important metrics that we investigate quantitatively using mathematical modelling.

The remainder of the article is structured as follows: background of DNA tile self-assembly, tile assembly models and tile pattern self-replication system are described in Sect. 2. Section 3 presents kinetic model of the tile pattern self-replicator. In Sect. 4, we present simulation results based on kinetic model, and discuss design choices in terms of assembly error and pattern size. Section 5, concludes the article.

2   Background

In this section we discuss briefly the background of the main concepts used in this article. This includes: a brief introduction to the self-assembly mechanism of DNA tile patterns, the abstract and kinetic modelling of tile assembly, and previously introduced tile pattern self-replication system [6].

2.1   Self-assembly of Programmable DNA Tile Patterns

A connection between algorithmic self-assembly and computation was studied by Wang in his theoretical tiling model [17]. The Wang Tiling theory demonstrates the implementation of a Turing machine by a finite set of square tiles with four colored edges.

Erik Winfree [20] applied the theoretical concepts of Wang tilling for the realization of programmable self-assembly patterns using DNA molecular structures (DNA tiles [19]) as analogue to the Wang's abstract tiles. DNA tiles serve as building blocks of self-assembly for the construction of 2-D physical patterns. DNA tiles consist of four ($\approx$ 50 nucleotide) ss-DNA molecules, synthesized for a given DNA tile design. Figure 2 illustrates the construction of a Double Crossover (DX) molecular DNA tile with four DNA strands. As shown in Fig. 2(a), each ss-DNA consists of a sequence of nucleotides (A, T, G, C). The tiles self-assemble through the bonding of these ss-DNAs at room temperature. The bonding process occurs when two complimentary strands meet and their

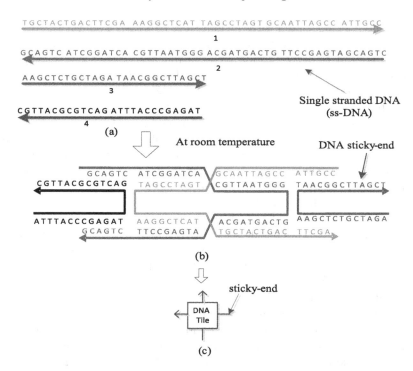

Fig. 2. DX DNA tile structure (a) Four ss-DNA (b) Assembled DNA tile (c) Abstract representation

base pairs: A-T and G-C, bind. Any left-over bases from each of the bonded strands form a sticky end(s) — as shown in Fig. 2(b). As the term implies, this end is available to "stick" or bond to another strand. DX molecular DNA tiles are square shaped structures where sticky-ends are represented by their respective square edges — as illustrated in Fig. 2(c).

2.2   Tile Self-assembly Models

The physical implementation of tile self-assembly in a wet-lab is often time-consuming, expensive and challenging with respect to reproducibility of results. Simulation of realistic models of DNA self-assembly provides a cheaper, faster (and more reliable) media in which to explore and refine new avenues of research, prior to experimentation. There are two simulation models of tile self-assembly, developed by Winfree [13,20]: (1) The abstract Tile Assembly Model ($aTAM$), and (2) The kinetic Tile Assembly Model ($kTAM$).

Abstract Tile Assembly Model (aTAM): The $aTAM$ [20] is based on Wang's tiling theory [17], which requires creation of a finite set of square shape tiles that are abstract representations of DX DNA-tile shown in Fig. 3(a). In aTAM, a tile t is represented by a quadruple $(\sigma_S(t), \sigma_W(t), \sigma_N(t), \sigma_E(t))$, where

$\sigma \in \Sigma$ is glue type associated with the four sides (North(N), South(S), West(W), East(E)) of a rotationally asymmetric unit square. The glue type, Σ, is a finite set, which is used to derive a glue strength function ($s : \Sigma \times \Sigma \rightarrow N$) for a legitimate tile association between two glues of tiles. The glue strength function is symmetric, i.e., $s(\sigma_1, \sigma_2) = s(\sigma_2, \sigma_1) \, \forall \sigma_1, \sigma_2 \in \Sigma$.

A tile pattern assembly system (TPAS) $\mathcal{T} = (T, S, s, \tau)$ consists of a finite set T of tile types, an assembly S termed as seed assembly, a glue strength function s and a temperature parameter $\tau \in Z^+$. A tile assembly system has a temperature 'τ' if any larger structure of tiles cannot be dissociated into smaller assemblies without breaking bonds of total strength at least 'τ'. Alternatively, a tile can join the assembly as long as the sum of the strengths of the bonds that it makes with tiles already in the assembly is at least τ.

Figure 3 illustrates the self-assembly process of the Sierpinski pattern [12, 21] at temperature 2 ($\tau = 2$). The tile set comprises a seed tile, two boundary tiles and four rule tiles - see Fig. 3(a). Tile edges are marked by non-negative integers illustrating their respective glue strengths. The South and West glues of the tiles are designed as inputs and the North and East glues are outputs.

Tile pattern assembly in the aTAM starts from a given seed structure that nucleates the pattern formation which grows into a finite or infinite pattern as more tiles join - see Fig. 3(b). Tiles join by forming bonds with strength at least of τ. For example, in a $\tau = 2$ assembly, each tile that binds with the growing pattern of tiles needs an attachment of total binding strength ≥ 2. For a given TPAS, a pattern assembly P is said to be terminal, if no tile can be added further that satisfies the $\tau - stability$ criteria.

The aTAM has given insights to important theoretical aspects of the tile assembly systems [3, 13]: (1) what can or can't be self-assembled?, and (2) if something can be assembled, how efficient it could be?

The Kinetic Tile Assembly Model (kTAM): Tile binding in the tile self-assembly is a reversible physico-chemical process that has been modelled using the kTAM [20]. The rate of tile attachment at a binding site of an aggregate is directly proportional to the tile concentration. The concentration of each type of tile (except the seed tile) can be given by $e^{-G_{mc}}$, where G_{mc} is the decrease in entropy when a tile binds at a vacant site. Therefore, the forward reaction rate (r_f) can be given by $r_f = k_f e^{-G_{mc}}$ where k_f is the reaction rate constant. Similarly, the tile detachment process is controlled by the energy required to break any single tile-aggregate bond and denoted by G_{se}. The value of G_{se} depends on the sticky end length (s) and the temperature (T), where $G_{se} \approx (4000/T - 11)s$. The tile reverse reaction rate involving b tile bonds is given by $r_{r,b} = k_f e^{-bG_{se}}$.

A larger value of G_{mc} thus implies a lower tile concentration and consequently a slower forward reaction rate (or vice versa). Similarly, a larger value of G_{se} results in a slower detachment rate. The optimum growth rate with low error rates happens near thermodynamic equilibrium ($G_{mc} \approx 2G_{se}$) [21], and may be given by $r^* \approx r_f - r_{r,2}$ and $\varepsilon \approx e^{-G_{se}}$, respectively. Therefore, a relation between optimum growth rate and minimum error rate may be given by $r^* \approx \beta \varepsilon^2$ where,

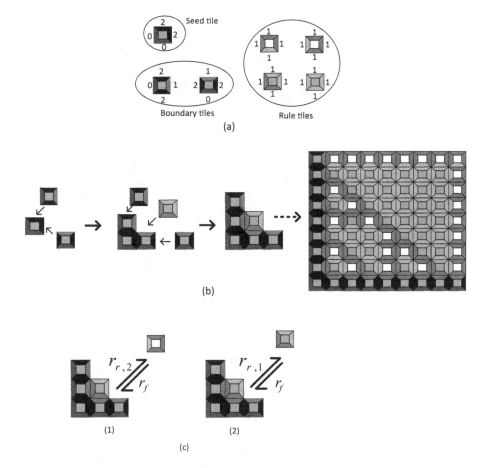

Fig. 3. Sierpinski pattern self-assembly. (a) Sierpinski tile set (XOR tile set). (b) Steps of self-assembly of Sierpinski pattern of size 9×9. (c) Kinetics of tile assembly in $kTAM$.

$\beta = 0.75 \times 10^6$ /M/sec. Thus, any effort to reduce the error rate (ε) by tuning physical parameters (G_{mc} and G_{se}) would result in a quadratic reduction of the growth rate. However, error reduction without significant fall off in assembly growth rate has been achieved by adding redundant tiles [8,21] and by protecting tile's inputs and outputs [5,11].

2.3 The Tile Pattern Self-replication System

Figure 4 shows the design of Tile Pattern Self-replication System (TPSS), earlier introduced in [6]. The L-shaped seed of the target pattern (P) is highlighted with a blue colour. The unique corner tile of the pattern is shown in red. Starting with the pattern structure (left cycle), pattern-mold (P-M) complex forms as CST attaches with the unique corner tile of the pattern, and further tiles from

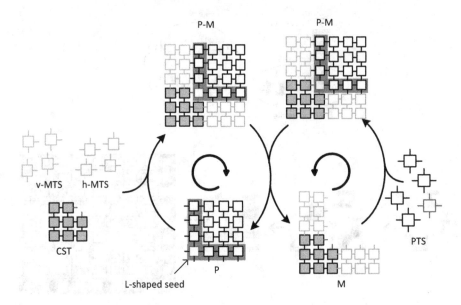

Fig. 4. Tile pattern self-replication system. (Color figure online)

the v-MTS and h-MTS sets assemble to form the vertical and horizontal arms of the mold, respectively. The P-M complex is dissociated into the seed and the mold (M) through external switching. The dissociated mold (M) serves as a new seed to assemble a new P-M complex (right cycle) that subsequently dissociates in the seed and the mold. Thus, the process initiates cross-coupled cycles catalyzing the formation of one another.

Let a pre-assembled target pattern, P, be self-replicated. L-shaped South-West border of the pattern P serves as seed, which enables entire rectangular pattern of tiles to be uniquely identified by the glues placed on its interior border. Considering that each tile in the pattern requires at least two bonds for a stable attachment (a case of $\tau = 2$ algorithmic tile self-assembly), the formation of pattern from the L-shaped seed would be a terminal assembly process [2]. A terminal assembly system forms a unique final structure from a set of supplied components.

The replication process starts with a pre-assembled rectangular pattern (P), Corner Super Tile (CST), and a set of Mold forming Tile Set (MTS). The MTS consists of two subsets: (1) Vertical Mold forming Tile Set (v-MTS) assembles to form a vertical double layer of the mold; (2) Horizontal Mold forming Tile Set(h-MTS) assembles the horizontal arm of the mold.

We require that the target pattern contains a unique, red-coloured tile on its lower-left corner position, which is not used on any other position inside the pattern. Observe that the CST consists of eight tiles, and therefore it is stable at temperature-2. The CST is designed to bind (using two strength-1 glues) on the special red-coloured tile. Mold formation is initiated with the binding of the

CST, and further proceeds as more tiles cooperatively join one by one until the entire South-West boundary of the pattern structure is covered by a double layer of tiles, creating a pattern-mold complex (*P-M*). Tiles forming the inner layer of the mold are designed as SWET type (now shown in the above schematics) with switch-enabled glue on the side that binds with the seed (pattern). The assembled pattern-mold complex undergoes a controlled dissociation, splitting into the Pattern *P* and the mold *M* structures. Observe that the dissociated mold structure has two layers of tiles, thus ensuring its stability under temperature-2 assembly framework.

In the next replication cycle, the dissociated pattern structure (*P*) repeats the left hand side pathway, and thereby, creates two (*P-M*) complexes, whereas the dissociated mold structure (*M*) drives the right hand side pathway using tiles from the PTS. Indeed, assuming we have at our disposal a tile set capable of assembling the pattern, we use the mold to reassemble the complete pattern P. Thus, by supplying the system with sufficiently many copies of the tiles within the MTS and PTS tile sets, and by continuing the process for i complete cycles, the replicator could theoretically produce 2^{i-1} copies of both the mold and the pattern structures. In a potential experimental implementation, one has to provide enough time for both the mold formation process (from a template pattern) and the pattern formation process (using the mold as a seed). Then, one adjusts the cycle of inhibitor signal supply, which triggers the pattern-mold dissociation such as to be at least as long as the maximum of the two expected time values.

3   Kinetic Model of the TPSS

In this section, we derive a simplified kinetic model of pattern self-replication system using the kTAM. The kinetic model consists of two cross-coupled pathways as shown in Fig. 5 — the left hand side pathway I corresponds to the cycle that is seeded with a target Pattern (P), whereas for the right hand side cycle *II*, Mold (M) acts as a seed. Intermediate product of both assembly pathways is Pattern-Mold (P-M) complex, which dissociates into copies of seed and mold.

Using the kTAM and its analytic model of kinetic trapping [20], macroscopic kinetic rates [1] (k_1 and k_2) of assembly steps leading to P-M complex from seed and mold are: $k_1 = \frac{r^*}{\sqrt{n^2+4}}$ and $k_2 = \frac{r^*}{n\sqrt{2}}$, respectively, where n denotes the size of $n \times n$ pattern and r^* is an optimal kinetic rate of tile assembly, as discussed in Sect. 2.2. Kinetic rate of dissociation of P-M complexes is a DNA strand displacement reaction. A typical kinetic rate of a toehold-mediated DNA strand displacement process involving a toehold of 3 nucleotides (nt), and a 7 nt long branch migration [22] is $k_d \approx 10^5 \ M^{-1}s^{-1}$.

Let number of copies of P, M and P-M structures at a time i are, $s[i]$, $m[i]$, and $sm[i]$, respectively. Therefore, under chemical equilibrium conditions, number of

---

[1] Macroscopic kinetic rate refers to an approximate kinetic rate for a terminal assembly process, as discussed in [4].

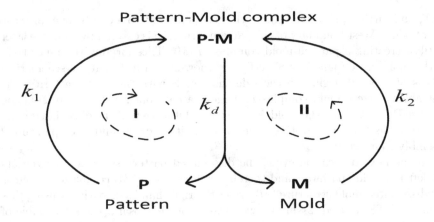

Fig. 5. Kinetic model of pattern self-replication system of rectangular patterns

copies of P, M, and P-M complexes in discrete time are given in Eqs. (1) and (2).

$$s[i+1] = m[I] = k_d sm[i] \tag{1}$$

$$sm[i] = k_1 s[i] + k_2 m[i-1] - k_d sm[i] \tag{2}$$

Using Eqs. (1) and (2), the concentration of pattern-mold complexes after i replication cycles is

$$sm[i] = \frac{k_1 s[i] + k_2 m[i-1]}{1 + k_d} \tag{3}$$

From Eqs. (3) and (1), $s[i+1]$ and $m[i]$ can be given as

$$s[i+1] = m[i] = \frac{k_d k_1}{1 + k_d} s[i] + \frac{k_d k_2}{1 + k_d} m[i-1] \tag{4}$$

The coefficient terms, $s[i]$ and $m[i-1]$ in Eq. (4), can be replaced with k_a and k_b, respectively. The equivalent equation is given as below

$$s[i+1] = k_a s[i] + k_b m[i-1] \tag{5}$$

Replacing $m[i-1]$ by $s[i-1]$ in Eq. (5) gives the following difference equations in discrete time.

$$s[i+1] = k_a s[i] + k_b s[i-1] \tag{6}$$

Applying z-transformation[2] in Eq. (6), it turns into following quadratic equation in the z-domain

$$z^2 - k_a z - k_b = 0 \tag{7}$$

---

[2] z-transform is a linear operator that is applied to convert non-linear difference equations of time (i) domain into linear equations of frequency domain.

Let λ_1 and λ_2 are the two roots of the quadratic equation (7): $\lambda_1 = \frac{k_a + \sqrt{k_a^2 + 4k_b}}{2}$ and $\lambda_2 = \frac{k_a - \sqrt{k_a^2 + 4k_b}}{2}$. Hence, a general solution of Eq. (5) can be represented as

$$s[i] = c_1 \lambda_1^i + c_2 \lambda_2^i \tag{8}$$

The c_1 and c_2 are arbitrary constants. For a replicator system supplied with c copies of seeds in the start, $s[0] = c$ and $m[0] = 0$, as mold is not yet assembled. Putting $i = 0$ in Eq. (5), it gives $s[1] = k_a c$. Applying these boundary conditions for $i = 0$ and $i = 1$ in (7), it gives

$$c = c_1 + c_2 \tag{9}$$

$$k_a c = c_1 \lambda_1 + c_2 \lambda_2 \tag{10}$$

Solving Eqs. (9) and (10) for c_1 and c_2, and putting these values in Eq. (8), the general solution of the difference equation (6) is

$$s[i] = \frac{(k_a - \lambda_2)c}{(\lambda_1 - \lambda_2)} \lambda_1^i - \frac{(k_a - \lambda_1)c}{(\lambda_1 - \lambda_2)} \lambda_2^i \tag{11}$$

The expression of $s[i]$ in Eq. (11) represents the population growth with respect to replication cycles (i). Clearly, the population $s[i]$ at a given replication cycle is proportional to the initially supplied seed concentration, and is a polynomial in λ_1 and λ_2. Dynamics of the population growth is governed by parameters: k_a, k_b, λ_1, and λ_2. These parameters depend on physicochemical conditions (local assembly temperature (T), tile concentration (G_{mc}) and total binding strength of tile (b and G_{se})) of the tile self-assembly medium.

4   Results and Discussion

The kTAM based analysis of tile assembly has demonstrated the effect of physical parameters over the growth rate and the error rate of tile assembly. It was established that a target error rate can be achieved only at a certain growth rate. Therefore, owing to the constraints of experimental feasibility, a proper choice for an optimum error rate and its corresponding growth rate has to be made. Herein, we analyse quantitatively the effect of these constraints over population growth dynamics and reliability of pattern replicator.

In Fig. 6, λ_1 is plotted against error rate and target pattern size using the replication gain derived in Eq. (11).
For an exponential gain of self-replication, value of λ_1 should be ≈ 2. From the plot, it is evident that for an error rate ($\epsilon \approx 5 \times 10^{-3}$), $\lambda_1 \approx 2$. In Fig. 7, λ_2 is plotted against error rate and target pattern size using mathematical model. From the plot, it is evident that $|\lambda_2| < 1$ for an error rate $\epsilon \approx 5 \times 10^{-3}$.

For a given set of kinetic parameter values of k_1, k_2 and k_d, $|\lambda_1|$ is > 1, and $|\lambda_2|$ is < 1. Therefore, an approximate population of self-replicating patterns after many replication cycles i.e., $i \to \infty$, can be given as

$$s[i] \approx \frac{(k_a - \lambda_2)c}{(\lambda_1 - \lambda_2)} \lambda_1^i \tag{12}$$

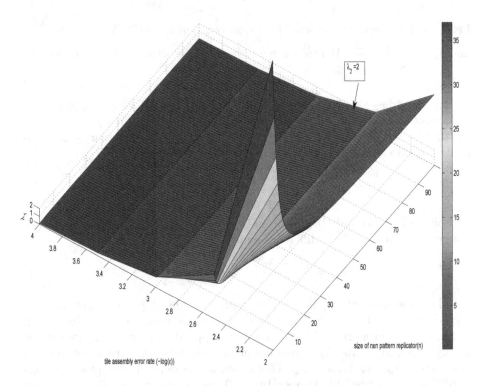

Fig. 6. λ_1 estimated from mathematical simulations

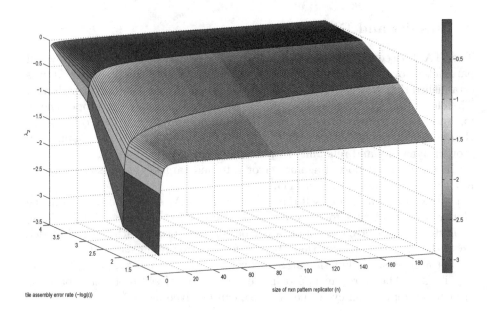

Fig. 7. λ_2 estimated from mathematical simulations

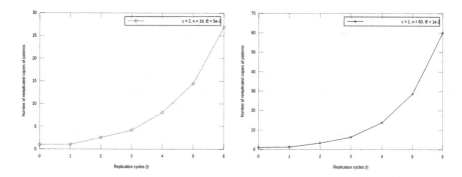

Fig. 8. Exponential replication growth of pattern self-replication: c = 2 n = 18, and ϵ = 5×10^{-3}) (LHS); c = 1, n = 60, and ϵ = 10^{-2} (RHS).

For an approximate replication gain, derived in Eq. (12), we plotted the pattern replication gain for replication cycles. Figure 8 shows exponential replication gains for two sets of parameters: initially introduced pre-assembled target patterns (c), size of target pattern (n), and assembly error rate (ϵ).

5   Conclusion

In this study, we constructed a kinetic model of a tile pattern self-replication system. Our model captures the dynamics of self-replicating tile patterns using equivalent kinetic rates for the two cross-coupled cycles of the self-replicator. The physico-chemical parameters of tile self-assembly influence the overall replication dynamics of the tile pattern self-replication process. It is observed that both size of target pattern and parameters should be carefully chosen so as to produce an exponential self-replication gain.

The observations of this paper could be useful for an experimental implementation of the pattern self-replication. To increase the robustness of self-replicating patterns in error accumulating tile self-assembly medium, error-correction tiles [7] can be used. A reliable self-replicator with error levels not exceeding a minimum threshold may further open up new directions for investigation of fundamental principles behind reproduction and selection-driven evolution.

References

1. Abel, Z., Benbernou, N., Damian, M., Demaine, E.D., Demaine, M.L., Flatland, R.Y., Kominers, S.D., Schweller, R.T.: Shape replication through self-assembly and RNase enzymes. In: SODA, pp. 1045–1064. SIAM (2010)
2. Adleman, L., Cheng, Q., Goel, A., Huang, M.D.: Running time and program size for self-assembled squares. In: Symposium on Theory of Computing (STOC), New York, pp. 740–748 (2001)

3. Barish, R.D., Rothemund, P.W., Winfree, E.: Two computational primitives for algorithmic self-assembly: copying and counting. Nano Lett. **5**(12), 2586–2592 (2005)
4. Czeizler, E., Popa, A.: Synthesizing minimal tile sets for complex patterns in the framework of patterned DNA self-assembly. Theor. Comput. Sci. **499**, 23–37 (2013)
5. Fujibayashi, K., Zhang, D.Y., Winfree, E., Murata, S.: Error suppression mechanisms for DNA tile self-assembly and their simulation. Nat. Comput. **8**(3), 589–612 (2009)
6. Gautam, V.K., Czeizler, E., Haddow, P.C., Kuiper, M.: Design of a minimal system for self-replication of rectangular patterns of DNA tiles. In: Dediu, A.-H., Lozano, M., Martín-Vide, C. (eds.) TPNC 2014. LNCS, vol. 8890, pp. 119–133. Springer, Cham (2014). doi:10.1007/978-3-319-13749-0_11
7. Gautam, V.K., Haddow, P.C., Kuiper, M.: Reliable self-assembly by self-triggered activation of enveloped DNA tiles. In: Dediu, A.-H., Martín-Vide, C., Truthe, B., Vega-Rodríguez, M.A. (eds.) TPNC 2013. LNCS, vol. 8273, pp. 68–79. Springer, Heidelberg (2013). doi:10.1007/978-3-642-45008-2_6
8. Chen, H.-L., Goel, A.: Error free self-assembly using error prone tiles. In: Ferretti, C., Mauri, G., Zandron, C. (eds.) DNA 2004. LNCS, vol. 3384, pp. 62–75. Springer, Heidelberg (2005). doi:10.1007/11493785_6
9. Keenan, A., Schweller, R.T., Zhong, X.: Exponential replication of patterns in the signal tile assembly model. In: Soloveichik, D., Yurke, B. (eds.) DNA 2013. LNCS, vol. 8141, pp. 118–132. Springer, Cham (2013). doi:10.1007/978-3-319-01928-4_9
10. von Kiedrowski, G.: A self-replicating hexadeoxynucleotide. Angew. Chem. Int. Ed. Engl. **25**(10), 932–935 (1986)
11. Majumder, U., LaBean, T.H., Reif, J.H.: Activatable tiles: compact, robust programmable assembly and other applications. In: Garzon, M.H., Yan, H. (eds.) DNA 2007. LNCS, vol. 4848, pp. 15–25. Springer, Heidelberg (2008). doi:10.1007/978-3-540-77962-9_2
12. Rothemund, P.W., Papadakis, N., Winfree, E.: Algorithmic self-assembly of DNA Sierpinski triangles. PLoS Biol. **2**(12), e424 (2004)
13. Rothemund, P.W.K., Winfree, E.: The program-size complexity of self-assembled squares. In: Proceedings of the Thirty-Second Annual ACM Symposium on Theory of Computing, STOC 2000, pp. 459–468. ACM (2000)
14. Schulman, R., Winfree, E.: Self-replication and evolution of DNA crystals. In: Capcarrère, M.S., Freitas, A.A., Bentley, P.J., Johnson, C.G., Timmis, J. (eds.) ECAL 2005. LNCS (LNAI), vol. 3630, pp. 734–743. Springer, Heidelberg (2005). doi:10.1007/11553090_74
15. Schulman, R., Yurke, B., Winfree, E.: Robust self-replication of combinatorial information via crystal growth and scission. Proc. Natl. Acad. Sci. **109**(17), 6405–6410 (2012)
16. Szathmry, E., Gladkih, I.: Sub-exponential growth and coexistence of non-enzymatically replicating templates. J. Theor. Biol. **138**(1), 55–58 (1989)
17. Wang, H.: Proving theorems by pattern recognition II. Bell Syst. Tech. J. **40**, 1–42 (1961)
18. Wang, T., Sha, R., Dreyfus, R., Leunissen, M.E., Maass, C., Pine, D.J., Chaikin, P.M., Seeman, N.C.: Self-replication of information-bearing nanoscale patterns. Nature **478**(7368), 225–228 (2011)
19. Winfree, E., Liu, F., Wenzler, L.A., Seeman, N.C.: Design and self-assembly of two-dimensional DNA crystals. Nature **394**(6693), 539–544 (1998)
20. Winfree, E.: Algorithmic Self-Assembly of DNA. Ph. D. thesis, California Institute of Technology Pasadena, California, USA (1998)

21. Winfree, E., Bekbolatov, R.: Proofreading tile sets: error correction for algorithmic self-assembly. In: Chen, J., Reif, J. (eds.) DNA 2003. LNCS, vol. 2943, pp. 126–144. Springer, Heidelberg (2004). doi:10.1007/978-3-540-24628-2_13
22. Zhang, D.Y., Turberfield, A.J., Yurke, B., Winfree, E.: Engineering entropy-driven reactions and networks catalyzed by DNA. Science **318**(5853), 1121–1125 (2007)
23. Zhang, D.Y., Yurke, B.: A DNA superstructure-based replicator without product inhibition. Nat. Comput. **5**(2), 183–202 (2006)

Retweet Influence on User Popularity Over Time: An Empirical Study

Yecely Aridaí Díaz-Beristain[✉], Guillermo-de-Jesús Hoyos-Rivera,
and Nicandro Cruz-Ramírez

Centro de Investigación en Inteligencia Artificial, Universidad Veracruzana,
Sebastián Camacho no 5, 91000 Xalapa, Veracruz, Mexico
yecely.diaz@gmail.com, {ghoyos,ncruz}@uv.mx

Abstract. Web-based Social Networks (W-bSN) have experienced a significant raise in terms of users, as well as the number of relationships among them. One crucial factor for this is the level of influence that a given user can have on other users, and how relationships emerge and disappear among users given the interest generated in a certain community by the posted commentaries. Twitter is the clearest case of W-bSN in which the relevance of the commentaries posted influences the way users create new relationships. In this paper, we analyze the cross influence among users, based on their area of interest, and the messages they post, and how relevant are these messages in the creation of new relationships.

Keywords: Social Network · Twitter · Data visualization influence

1 Introduction

Web-based Social Networks (W-bSN's) have enormously evolved since their dawn in the late 90's. Many of them had a relative success, such as Microsoft©'s MySpace® or Hi5®. They established the basis for developing a new generation of W-bSN's. Recently, there is a growing interest in two of the most used: Twitter® and Facebook®. Each of them has a considerable amount of users worldwide, and their activity represents many terabytes of information exchanged every day.

Twitter is considered to be a special case of W-bSN, called "micro-blogging". It implements an asymmetrical model, since there is no need of an authorization in order, for a given user, to create a new relationship (*follow*) with another user. A user willing to receive updates of the activities of another user, e.g. the tweets he posts, only needs to *follow* him, with no action required by this latter. Users' main activity in Twitter is posting *tweets*, short messages of at most 140 characters long, expressing thoughts, ideas, feelings or opinions. Once a user tweets a message, all the following users are notified of this. Another action that can be performed in Twitter is the, so-called, *retweet* (RT), which corresponds to the action of spreading a message previously twitted or RT by another user. It is important to note that the retweeting users do not necessarily have to be following a given user in order for this to be able to retweet him.

Consequently, this model produces some tweets which, given their relevance, are widely spread, becoming eventually viral and, using the Twitter terminology, become

© Springer International Publishing AG 2017
R. Prasath and A. Gelbukh (Eds.): MIKE 2016, LNAI 10089, pp. 38–48, 2017.
DOI: 10.1007/978-3-319-58130-9_4

a *Trending Topic* (TT). Thus, a TT is a topic a lot of people is interested about and, in consequence, decides to spread it. Given these characteristics, Twitter has emerged as a suitable platform through which people can try to become popular, i.e., have more followers everyday. If this is the utterly intention, a given user will try to tweet on interesting or controversial subjects, so that people can get interested in them, and eventually can become his follower. In addition, there are passive users who only wish to keep informed about tweets of the people they follow, and do not have an active presence in Twitter. In this order of ideas, it becomes a subject of study to define the factors that can, eventually, allow a user to get a follower of someone else. A user with an important number of followers can be considered as an influential one, since each idea he expresses will automatically reach a big number of users of the platform.

In this paper, we present results of our work, which consisted in analyzing, the factors that may lead that a user gets more followers, and in this way, eventually become influential.

This paper is organized as follows. In Sect. 2 we describe the related work and what other authors have proposed. In Sect. 3 we present the basis for our experimentation. In Sect. 4 the elements necessary to initiate our experimentation. In Sect. 5 the results of our experiment. In Sect. 6 we draw some conclusions and propose some directions for future research. Finally, we present the references.

2 Related Work

Since its creation in 2006, Twitter has gained notoriety and popularity. Currently, it has about 313 million active users every month. It has become a near real-time information spreading way widely used worldwide, and its relevance has enormously growth. As explained before, some users tend to have more followers due to the fact that other users get interested in their opinions. These users are considered as influential ones. The relevant element to analyze here is knowing which factors can be considered as important so as to confer a user more followers, and how it influences the number of followers of other users. There are several works that study the influential or "valuable" users [1, 2], the impact of tweeting and retweeting [3, 4], viral marketing [5], etc., which are used to understand the spread of information and the level of the user influence.

Cha et al. [6] present an empirical study of the patterns of influence on users considered as popular. They study three main factors: in-degree (the numbers of users following the user under analysis), number of RTs, and number of mentions. They propose the existence of influential users, i.e., those who can make their tweets to be widely retweeted, as well as to receive a big amount of mentions. From their analysis they concluded that such sort of users tend to publish tweets on controversial subjects. Also, their study affirms that users who limit their tweets to a single topic show a greater increase in the level of influence.

Romero et al. [7] realize the intuitive idea of some users are harder to influence because they are not interested in creating or sharing information, argue that the most of Twitter users are passive. According with the author's passivity is a barrier to propagation, while some users retweet a lot, others do not do it very often. They propose an algorithm similar to Hyperlink-Induced Topic Search (HITS) and PageRank to measure the influence considering not only the number of followers, but also the RTs and mentions. They found that influential users are highly active, and as a consequence defined a new influence measure based on user activity.

Retweeting, as stated in [8], has a preponderant importance since the fact of executing such action indicates not only interest in a given tweet, but also the level of confidence deposited in the original publisher, as well as the agreement with the content. This is an important conclusion that helps us supporting our work later.

Another case, presented in [9], affirms that the propagation of the tweets tends to happen by users that have shown to be influential in the past, and who also have an important number of followers. They propose a formula allowing to measure the influence of users taking in consideration the number of RTs and number of mentions they have.

Other researchers have focused in proposing methods to measure the influence of users in subjects with similar topics. Anagnostopoulos et al. [10] define the level of influence as the fact that one individual can induce another individual to act in a similar way. Such type of users are called to be "active". They present a probabilistic model that evaluates when a user becomes active in a period of time, and they assume that their friends (this is how the author refers to the followers) increase their own probability of becoming active too. They concluded that people are influencing each other every discrete time and estimated the maximum likelihood.

In [11] Crandall et al. study the influence of users based on the "homophily". This term refers to the level of similarity of people that interact with each other. They divide it in social influence and selection. The first is when people pick up behaviors related with people they interact with, and the selection is when they seek out for similar users to interact with. They quantify the similarity of users over time consider the topic of interest of each user. The authors proposed a model of user behavior where individual users can interact with others and then select the users with a higher number of activities and interactions referred as to influential users.

The work of Weng et al. [12] consists in identifying influential users on Twitter. The strategy is similar to the PageRank algorithm. Their algorithm considers topics extracted from tweets. One of the main contributions is that they compute each user's topic distribution based on their tweets using LDA, showing that topics of connected users are significantly correlated.

Compared to the previously presented works, ours can be considered as an experimental framework allowing us to analyze the real impact, using real data, that tweets, RTs, and mentions, may have in the level of popularity of Twitter users. We assume as our basic hypothesis that RTs and mentions made by influential users have an effect on the number of followers of a given user.

3   Basic Experiment Framework

The influence is the ability that an individual has to modify the perception or beliefs of other people. The reputation of a user has a direct effect on the perception and opinions of other individuals, and can be effectively used to obtain advantages. The opinion expressed by an influential user can produce, as an effect, that other users change their mind about what they previously thought. For a long time, there have been studies in fields such as sociology, politics, and marketing about the influence that experts, or recognized people, may have, and in consequence, to understand why certain trends appear. For example a campaign turns to be more effective if a message related to it becomes viral. The theory of the traditional communication [13, 14] affirms that a minority of people belonging to a group, denominated distinguished influential, become natural leaders with the capacity of persuading others.

In W-bSN, particularly Twitter, we can also find outstanding users in certain fields and with a number of followers who maintain a more or less permanent interest in their opinions. These users become participatory entities, mentioning a user they are interested on, and sharing those ideas or opinions with their respective followers.

Our main interest consists in analyzing the influence patterns among Twitter users, and how the users considered as experts in a given field can promote the growth of the number of followers of other users, positioning this last through the use of RTs.

Taking into account the proposal of [15], we assumed 18 thematic categories of main subjects of interest (art and design, books, business, charity & deals, fashion, food and drinks, health, holidays & dates, humor, music, politics, religion, science, sports, technology, TV & movies, other news, other). Then we defined six linguistics values based on the number of followers of a given user, as shown in Table 1.

Table 1. Linguistic values and their range

Linguistics values	Range of followers	Figure
Unknown	0 to 1,000	Circle
Ordinary	1,001 to 10,000	Square
Outstanding 1	10,001 to 100,000	Triangle
Outstanding 2	100,001 to 1,000,000	Diamond
Outstanding 3	1,000,001 to 10,000,000	Pentagon
Famous	10,000,001 and more	Heptagon

As we can see on Table 1, the "Unknown" linguistic value represents users who either, have just created their accounts, or users with very little activity, generating null attraction for other users. The "Ordinary" users are those who start gaining some popularity and, in consequence, start having new followers who are interested in their timeline and RT him. For the "Outstanding" users, we have defined three different levels based on their number of followers, representing active user accounts that realize diverse posts during the day, so that still more users decide to follow them. According to our results we could infer that these kinds of users have opinions that are respected, so that their influence can be known as important. Finally, for "Famous" users we consider those users who have a huge number of followers. Some users of this type are: @katyperry (92,044,564), @justinbieber (86,775,392), and @BarackObama (76,878,181), to name some.

We found no information about the number of users belonging to each linguistic value. So, we obtained a sample of a million random users to figure out the proportions of the accounts: 98% correspond to the "Unknown" linguistic value, 1.53% to "Ordinary", "Outstanding 1" is has 0.35%, "Outstanding 2" the 0.072%, the "Outstanding 3" has 0.040% and just the 0.008% correspond to "Famous".

4   Experiment Design

In this section, we present the design of the empirical study to analyze the importance of patterns through Twitter users and how they can endorse the increase the number of followers some other users have, by means of the use of RTs.

As the first part, we picked a user to be the subject of our analysis, which we will call in the following to be the *Root* user. Using the Twitter API we extracted all the information about the activity of *Root*, that is, tweets, RTs, mentions and new followers. At the beginning of the observation, *Root* had a total of 3,253 followers and his tweets were classified mainly as belonging to in the Technology category. Since the creation of his Twitter account, the growth of followers had a relative stable behavior, getting at most two or three new followers per week, also associated to the Technology category.

Then his behavior was modified through the diversification of his main publishing interests, as well as the use of additional resources to plain text, such as embedded images/videos, or URLs. The new categories in which this user newly participated included Sports and Music. We must explicitly mention that, even if this user used to publish mainly about Technology, tweets posted in other areas contained no relevant information, so other users generally ignored them. In this context, tweets having these other categories as the main subject, were analyzed by the algorithm of [15]. Once we were sure that they corresponded to the desired category, we included a commonly used hashtag (HT) in the tweet. This was made with the intention of making his tweets visible in currently existing conversation threads.

5   Experiment Results

In our experiment we firstly illustrate in Fig. 1 the fact that, there was a regular behavior of *Root* during the seven previous weeks (t_0) until the first phase of the experiment (t_0') corresponding to four weeks. Then, we increasing the amount of tweets (t_0'') and diversified the subjects (Sports, Politics, Businesses and Others categories). As it can be clearly observed, the amount of new followers per unit of time (week) grows at the right side of the figure.

In Fig. 2. we present a zoom corresponding to the period between t_0' and t_0'', which corresponds to the period of fourteen weeks that lasted our experiment. As it can be seen, the number of new followers increased when *Root* published a tweet in the alternative categories. In this figure P correspond to Politics, T to Technology, B for business and O for others. Vertical lines headed by the letter of the chosen category represents each time *Root* posted a new tweet on a specific category, seeing then the amount of new followers associated to that tweet. It can be observed that tweets of the Politics

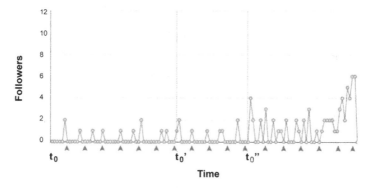

Fig. 1. Regular (t_0), slight (t_0') and strong behavior (t_0'')

category obtained a substantial increase in the number of followers, because users with "Outstanding 1" linguistic value retweeted the original tweet, and their followers considered it as relevant. The horizontal line at the bottom of Fig. 2 corresponds to part illustrated in Fig. 3, where we can observe the correlation that exists between the RT action and the follow action.

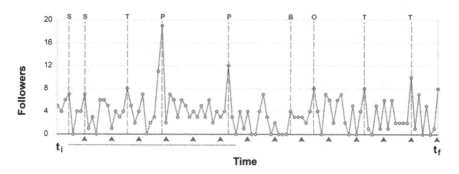

Fig. 2. The beginning (t_i) and the conclusion (t_f) of the accomplished experiment

In this context, we can see that each time *Root* tweets were retweeted (in gray), especially by users with great quantities of followers, his number of new followers increased.

As a collateral result of our experiment, it is important to note that not every user having retweeted a given *Root* tweet, decided to follow him. These users were reached through a RT by an intermediary user they follow, and then they considered the original opinion relevant enough to spread it, but in a first time they did not consider the *Root* user as interesting enough a to start a follow relation on him. Of all the users who did a RT only 37.5% decided to start following. It is important to mention that some users began following *Root* and at the end the 3.2% decided to retire the relation to *Root* user.

In Fig. 4 we present the users that started following *Root* after a RT of a *Root's* tweet. By using different symbols, we represent the linguistic values of the other types of users. The star *Root*. Here it can be observed that, per example, the "Outstanding 1" (triangle)

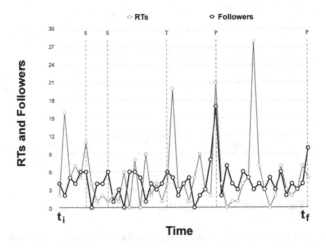

Fig. 3. Correlation that exists between RTs and follows

user produced more new followers for *Root* than those produced by an "Unknown" (circle) user. During the experiment corresponding to the diversification of subjects we also discovered that the inclusion of additional resources within a tweet and the determination of the optimal size of the HT, are factors that must also be taken into account in order to get the attention of other users.

The first of these strategies was the incorporation of additional content to the tweet; this can be expressed under the form of image, video or URL. In Fig. 5 we can observe that, the kind of resource included in the tweet, impacted the number of RTs. In this way, in order of relevance, it is possible to observe that the highest impact was obtained

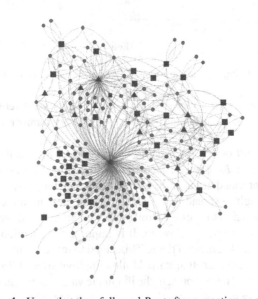

Fig. 4. Users that they followed *Root* after a mention or a RT

when a URL is attached, then when an image is associated, and finally when attaching a video. It calls our attention the fact that when a video is included the impact is minimum compared with the other alternatives. We can affirm that we agree with Zarella [16], where it was affirmed that the inclusion of images or URLs does more attractive a tweet for the users.

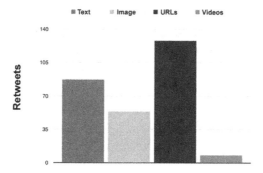

Fig. 5. Including additional content to the tweet

The use of HTs is another way to reach more Twitter users, but choosing an appropriate length for them is crucial for its success. In Fig. 6 we can see that HTs formed by a single word were more successful than those formed by two to four concatenated words.

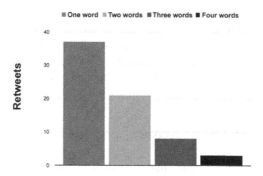

Fig. 6. Including a HT to tweet

In this sense, we disagree with Weng et al. [12] who affirmed that to include HTs it is not important, demonstrating that using a HT of proper size creates more interest in the user. Based on these preliminary it is possible to deduce that the simplicity is a key element. This is shown in Fig. 6.

In order to validate our results, we applied this same experiment to a set of twelve randomly selected, but assuring to have a representation of all our linguistic categories, during a period of one month. The results are shown in Table 2. What we could observe is that the results obtained here correspond with the results obtained for our *Root*. Users belonging to the "Unknown" or "Ordinary" categories tend to remain in these categories since their tweets are not very interesting or diverse. On the other hand, users belonging to the "Outstanding" or "Famous" linguistic values, who usually are diverse in the way they tweet, in the same sense as stated in this work, tend to get more users as time passes.

Table 2. Growth in number of followers

User	Linguistics values	New followers
a_1	Unknown	4
a_2	Unknown	−3
a_3	Ordinary	75
a_4	Ordinary	−14
a_5	Outstanding 1	383
a_6	Outstanding 1	865
a_7	Outstanding 2	5,371
a_8	Outstanding 2	49,692
a_9	Outstanding 3	26,940
a_{10}	Outstanding 3	7,727
a_{11}	Famous	100,424
a_{12}	Famous	532,222

In fact, another observation that we could make is that users tend to remain in the same linguistic category. In this sense, users categorized as "Famous" have an exponential growth in the number of followers; this is a logic consequence since, as more followers spread their tweets, it is easier to gain new followers. For example, the account to @BarackObama in a period of two months had 1,417,820 new followers and tweets with at least 400 RTs. And although some users stopped following him, the number of new followers was bigger than the number of users who stopped following him.

6   Conclusions and Future Work

As results of our work we concluded that it was not easy to determine the factors that allow a Twitter user to get more followers. Firstly, we could observe that users tend to keep steady, and it is not very frequent that a user changes of linguistic category. However, we could demonstrate that there exist means through which a user wishing to get more followers, can get it.

The use of several tools, such as HTs, mentions, and included resources, shown to be a good way of getting more users interested in one's timeline. The length of the HT is another factor that has an influence. As shorter and simpler HTs are used, higher is their probability of success. Another factor to be taken into account is to get users of the "Outstanding" of "Famous" linguistic categories to get interested in one's tweets. If one of these users gets interested enough in what we tweet, he can retweet, and then our probabilities of success are higher. Finally, diversifying the subjects of discussion helps a lot also to get more followers.

The proposed strategies can be used on any type of account, not mattering if it is personal or corporative. Our research proposes guidelines to continue with the study of the influence of the users and criteria of growth.

The review field of study provides an opportunity for future research. We considered to focus in analyzing exclusive the behavior of "Outstanding 3" and "Famous" users and thus define new metrics and could be implemented in a model to describe the behavior of influencers.

References

1. Lee, C., Kwak, H., Park, H., Moon, S.: Finding influentials based on the temporal order of information adoption in twitter. In: Proceedings of the 19th International Conference on World Wide Web, pp. 1137–1138. ACM, April 2010. Maxwell, J.C.: A Treatise on Electricity and Magnetism, 3rd ed., vol. 2. Oxford, Clarendon, pp. 68–73 (1892)
2. Quercia, D., Ellis, J., Capra, L., Crowcroft, J.: In the mood for being influential on twitter. In: 2011 IEEE Third International Conference on Privacy, Security, Risk and Trust (PASSAT) and 2011 IEEE Third International Conference on Social Computing (SocialCom), pp. 307–314. IEEE, October 2011
3. Suh, B., Hong, L., Pirolli, P., Chi, E.H.: Want to be retweeted? large scale analytics on factors impacting retweet in twitter network. In: 2010 IEEE second international conference on Social computing (socialcom), pp. 177–184. IEEE, August 2010. Foster, I., Kesselman, C.: The Grid: Blueprint for a New Computing Infrastructure. Morgan Kaufmann, San Francisco (1999)
4. Boyd, D., Golder, S., Lotan, G.: Tweet, tweet, retweet: conversational aspects of retweeting on twitter. In: 2010 43rd Hawaii International Conference on System Sciences (HICSS), pp. 1–10. IEEE, January 2010. Foster, I., Kesselman, C., Nick, J., Tuecke, S.: The physiology of the grid: an open grid services architecture for distributed systems integration. Technical report, Global Grid Forum (2002)
5. Leskovec, J., Adamic, L.A., Huberman, B.A.: The dynamics of viral marketing. ACM Trans.Web (TWEB) 1, 5 (2007)
6. Cha, M., Haddadi, H., Benevenuto, F., Gummadi, P.K.: Measuring User Influence in Twitter: The Million Follower Fallacy. ICWSM 10(10–17), 30 (2010)
7. Romero, D.M., Galuba, W., Asur, S., Huberman, Bernardo A.: Influence and passivity in social media. In: Gunopulos, D., Hofmann, T., Malerba, D., Vazirgiannis, M. (eds.) ECML PKDD 2011. LNCS (LNAI), vol. 6913, pp. 18–33. Springer, Heidelberg (2011). doi: 10.1007/978-3-642-23808-6_2
8. Metaxas, P., Mustafaraj, E., Wong, K., Zeng, L., O'Keefe, M., Finn, S.: What do retweets indicate? results from user survey and meta-review of research. In: Ninth International AAAI Conference on Web and Social Media, April 2015
9. Bakshy, E., Hofman, J.M., Mason, W.A., Watts, D.J.: Everyone's an influencer: quantifying influence on twitter. In: Proceedings of the Fourth ACM International Conference on Web Search and Data Mining, pp. 65–74. ACM, February 2011
10. Anagnostopoulos, A., Kumar, R., Mahdian, M.: Influence and correlation in social networks. In: Proceedings of the 14th ACM SIGKDD International Conference on Knowledge Discovery and Data Mining, pp. 7–15. ACM, August 2008

11. Crandall, D., Cosley, D., Huttenlocher, D., Kleinberg, J., Suri, S.: Feedback effects between similarity and social influence in online communities. In: Proceedings of the 14th ACM SIGKDD International Conference on Knowledge Discovery and Data Mining, pp. 160–168. ACM, August 2008
12. Weng, J., Lim, E. P., Jiang, J., He, Q.: Twitterrank: finding topic-sensitive influential twitterers. In: Proceedings of the Third ACM International Conference on Web Search and Data Mining, pp. 261–270. ACM, February 2010
13. Rogers, E.M.: Diffusion of Innovations. Simon and Schuster, New York (2010)
14. Miller, G.R., Burgoon, M.: Persuasion research: review and commentary. Commun. Yearb. **2**, 29–47 (1978)
15. Lee, K., Palsetia, D., Narayanan, R., Patwary, M.M.A., Agrawal, A., Choudhary, A.: Twitter trending topic classification. In: 2011 IEEE 11th International Conference on Data Mining Workshops (ICDMW), pp. 251–258. IEEE, December 2011
16. Zarrella, D.: The science of retweets. Accessed 15 December 2009

Privacy Preserving Interceptor for Online Social Media Applications

T. Shanmughapria$^{(\boxtimes)}$ and S. Swamynathan

Department of Information Science and Technology, Anna University, Chennai, India
priyanethiran@gmail.com, swamyns@annauniv.edu

Abstract. In the current scenario, online social network (OSN) plays a vital role in user's day to day life. User's store lots of their personal information in OSN, and they share their day to day experiences with their connections. There are enormous number of third party application (TPA) services allied with OSN to provide extended services to the users. The OSN manages the identity verification by authenticating the user and granting access to allied TPAs. The TPA requests permissions to access personal attributes about the user when accessed by the user for the first time. The personal attributes marked as required by the TPA has to be shared to avail the service. The privacy risk increases exponentially with the users TPA usage. The users not only leak their information to TPA but also end up unlocking a new type of threat from correlation with auxiliary information, through the data available from alternative sources. In this paper, we focus on reducing the level of sensitive data exposed to the external parties. The feasibility of providing such a service by restricting the data flow through access control policies is not feasible with the current All or Nothing approach. Hence, in this paper, we propose, the Privacy Preserving Interceptor (PPI) that acts as an interceptor between OSN and TPA to provide the required utility and yet preserves user's privacy. PPI identifies the sensitive attributes shared by the user and transforms the original data into a less sensitive form that still meets the utility goals. Standard Differential Privacy in combination with other perturbation mechanism of replacing with random values is used in PPI. The users privacy remains more or less the same both before and after the data share to TPA and still meets the utility needs of the user.

Keywords: Online social networks · Third party applications · Privacy · Data perturbation · Differential privacy

1 Introduction

To capture the interactions between the user, OSN, and TPA, we have chosen the Facebook platform [1] for observation. When the users access a TPA for the first time, the users are presented with a list of attributes to be shared with the TPA. The users are allowed to access the TPA only if they agree to share

© Springer International Publishing AG 2017
R. Prasath and A. Gelbukh (Eds.): MIKE 2016, LNAI 10089, pp. 49–57, 2017.
DOI: 10.1007/978-3-319-58130-9_5

the attributes. Otherwise, the service is restricted. Though, such data could be used constructively for sharing users experience and use that knowledge to improve the experience of other users availing the same service; the users are being exposed to more and more privacy threats in such situations. The very act of sharing data is not considered as a breach of privacy. The privacy breach occurs whenever the contextual integrity is wrecked, in other words whenever the information not intended to be disclosed in that particular context gets exposed.

Risk Analysis. The risk involved in sharing attributes may lead to attacks [3,5,10,13] like privacy violations, de-anonymization, fake profile creation, information leakage attacks, cyber bullying, identity theft, etc., Women and children are even more vulnerable to these kinds of attacks. [2,15,17]. In an online survey conducted by BullyingUK, it is recorded that 87% of teens of age between 11 and 16, who reported cyber abuse said they were targeted on Facebook, and 20% blamed Twitter [14]. There were reported incidents where social media applications violated the accepted terms of service [9,11]. Consider a scenario where the user shares their attributes (sensitive or insensitive) to many applications from the same application developer. The application developer has access to all the information about a user from all the applications developed by them. There is a possibility to correlate that information and other information available through online resources, to infer the information that was not intended to be shared, thus violating the contextual integrity.

Motivation. Applying access control does not solve the current issue of sharing user attributes. Restricting sharing of a required attribute to a TPA, restricts the user accessing the service. Hence a solution is needed to enable the user to share all the required variables in a privacy-preserving way. PPI applies data perturbation techniques like standard differential privacy and randomization to transform the data. As the Privacy Preserving Interceptor resides in between OSN and TPA, users will benefit by using more number of TPA without much privacy concern and on the other hand, OSN and TPA will have an increased number of active users participating. Considering all these benefits, there should be some mutual consensus set up between OSN, PPI, and TPA for such interactions to be feasible. The rest of the paper is organized as follows. Section 2 presents the design of privacy preserving interceptor. Section 3 presents a survey of the existing work in the literature and Sect. 4 presents conclusion and future extensions of the paper.

2   Privacy Preserving Interceptor

2.1   Overview

The goal of the proposed mechanism is to preserve the privacy of the user and enable users to access multiple TPA without much compromise to utility. PPI applies Standard Differential Privacy concepts, Where the disclosure risk of the user remains the same as before sharing. Utility level of the data is managed as we share all the required attributes as perturbed variables. The perturbed variable resembles the original variable statistically to provide the necessary level of utility. The degree of perturbation depends on the user's privacy requirement. Hence, our design provides a customized solution to the user that strikes a balance between privacy and utility matching the user's need.

The overall interaction between the components OSN, TPA and PPI is shown in the Fig. 1. The users request to OSN is always intercepted by PPI, it identifies the attributes and applies perturbation matching the user's level of privacy. The user's level of privacy is captured based on the user's sharing behavior captured via PPI. Initially, for a new user the privacy level is measured based on a short survey conducted when the user starts to use the application. The PPI eventually learns about the user's actual sharing level based on the attributes shared by users.

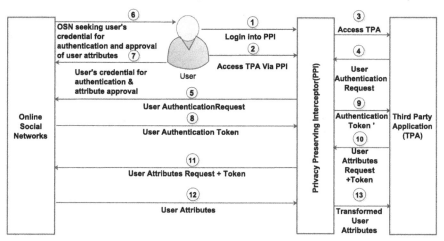

Fig. 1. OSN, PPI, TPA interaction pattern

2.2   Design of PPI

Differential Privacy. A randomized function f gives $\in$ differential privacy if for all data sets D_1 and D_2 differing on a single user and all S $\subseteq$ Range(f).

$$Pr(f(D_1) \in S) \le e^{\in} \quad Pr(f(D_2) \in S) \tag{1}$$

where $f(D_1)$ is function applied to data set including the user attributes and $f(D_2)$ is function applied to data set excluding the user attributes and $\in$ gives the required privacy level for all user attributes in the set S. The value of $\in$ is chosen based on the factors deciding the privacy requirement like user's privacy choice and attributes sensitivity.

Laplacian noise as in Eq. 2 is generated with chosen $\in$ and added to original user attributes to generate the perturbed user attributes.

$$\delta_{noise} = \frac{1}{2b} \quad e^{\frac{-(x-\mu)}{b}} \tag{2}$$

where μ is the mean of the noise signal and b represents the spread of the noise. The spread parameter is based on the global sensitivity parameter $\Delta F(x)$ and the privacy parameter $\in$ as shown in Eq. 3.

$$b = \frac{\Delta F(x)}{\in} \tag{3}$$

The global sensitivity is the difference between the maximum value and the minimum value that could be assigned to an attribute as in Eq. 4.

$$\Delta F(x) = max - min \tag{4}$$

The attribute requested by TPA is replaced with modified value $Attr_{perturbed}$ as in Eq. 5 arrived by adding the laplace noise δ_{noise} to original attribute (Fig. 2).

$$Attr_{perturbed} = Attr_{original} + \delta_{noise} \tag{5}$$

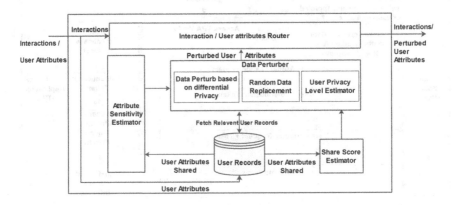

Fig. 2. PPI block diagram

Choosing the Privacy Parameter. Individual's choice of privacy varies based on multiple constraints like their geographic location, community, age, gender, etc. Therefore, there cannot be a single fixed level of privacy for all users. To address this issue, the module User's privacy level estimator updates the sharing level of the user based on the user's sharing behavior. Initially, a survey was conducted from users to capture their desired sharing level. The desired sharing level varies with the actual sharing level based on the context in which it is shared. So, the share score estimator calculates the actual sharing level of the user from the attributes shared in OSN. The privacy level of the user is measured on a scale of 3 based on the survey questions and the attributes shared by the user in OSN. The attribute table shown in Fig. 3 is used to calculate the share score as in Eq. 6.

$$Sharescore(actual) = \frac{w_1 \Sigma att_1 + w_2 \Sigma att_2 + ... + w_n \Sigma att_n}{n * m} \tag{6}$$

where $w_1, w_2...w_n$ are the weights assigned to the attributes, $\Sigma att_1, \Sigma att_2...$ Σatt_n are attributes, n is the number of attributes and m is the number of applications. There may be variations between actual and desired share score. So, an average of actual and desired recorded from survey questions is used to measure the Share score. The attribute sensitivity estimator finds the sensitivity of data. The sensitivity of the data attributes is measured by finding how frequently the attribute has been shared (the most sensitive attribute is the least shared attribute) and the default classification given by Facebook. The attribute accessed by users are stored in structure as shown in Fig. 3.

Attribute/ Application	Attribute 1	Attribute 2	Attribute 3	...	Attribute N
Application 1	1	0	1		1
Application 2	1	0	0		1
Application 3	1	0	1		0
...					
Application M	0	1	1		1

Fig. 3. Attribute table

Value '1' stored in the table denotes the attribute being shared. Facebook's classification is shown in Fig. 4, the attribute is assigned weight factor 1 for basic profile elements, weight factor 2 for extended profile elements and a weight factor of 3 for extended permissions. Let N be the number of applications accessed by the user. The sensitivity value calculation for attributes is done as in Eq. 7.

$$sensitivityscore_{att} = \frac{\sum_{i=1}^{N} app_i * wt_{classification}}{\sum_{i=1}^{N} app_i} \quad (7)$$

finally privacy level $\in$ is calculated as in Eq. 8.

$$\in = Sensitivityscore_{att} + Sharescore \quad (8)$$

Replacement with Random Values. Certain attributes like name, email id when shared with TPA, the chance is high that it could be correlated with auxiliary information revealing much more information, than currently available. Instead of adding noise to it, PPI uses the approach of distorting the original values by replacing with random attributes. Based on the user's sharing choice the attributes are either shared or replaced with random values picked from the database.

Permissions / Attributes	Basic	Extended Profile Properties	Extended Permissions
Public Profile	*		
FriendsList	*		
Email	*		
Personal Description		*	
Work History		*	
Education History		*	
Home Town		*	
Birthday		*	
Likes		*	
Current City		*	
Friends Birthday			*
Messages			*
Checkins			*
Friends Personal Info			*
Relationships			*
Photos			*
Relationship Status			*

Fig. 4. Attribute classification

2.3 Discussion

The optional attributes can be opted not to share and the required attributes are perturbed thereby balancing the privacy and utility requirement of the user. Let us consider a case for App - Livestream, the attributes requested are public profile elements (name, age, gender, picture), email. Sensitivity value calculated for public profile elements is 1, email is 2. privacy level $\in$ is calculated as 1.22 and laplacian noise (0, 8.19) i.e., 0 mean and spread of 8.19 is added to original value. Assuming data ('AAA',23,'f',pic1.jpg,xyz@gmail.com) is transformed as data' ('XYZ',28,'f',pic1.jpg,xyz@gmail.com). The user can also opt to change the profile photo and email to be replaced with some random elements. PPI provides an attempt to perturb the required attributes. The mechanism could be applied to birthdate, location parameter, timezone, age etc., and random replacement has been applied for all other attributes. In our future extention we wish to apply other techniques to identify the similarity and replace the attributes like likes, action.music, action.books etc., with similar items to preserve the utility. The trust worthiness of the TPA could also be included as a parameter in deciding how much to disclose. Privacy Level Vs Perturbation % is plotted in Fig. 5. value of $\in = 1$ provides the highest perturbation level with 23% noise and value of $\in = 6$ provides the lowest perturbation with 4.5% noise.

3   Related Work-Comparison

There has been considerable work done to resolve the privacy issues in OSN-TPA scenario. To understand the seriousness of the privacy risk involved in

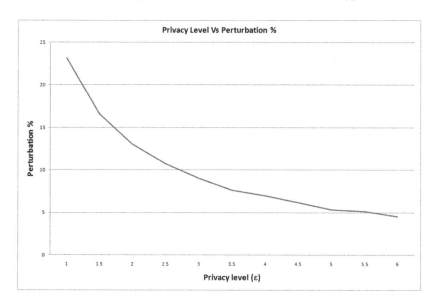

Fig. 5. Privacy level vs perturbation %

sharing user attributes with TPA Chaabane et al. [6] conducted an experiment to study the interaction between OSN applications and other external parties. The research revealed shocking results that the Facebook and RenRen applications interacted with hundreds of different fourth-party tracking entities. A similar study was conducted by Aldhafferi et al. [3] showed that the Personal data collected through TPA could be used for data matching to reveal sensitive information posing serious risks to privacy. [12] Kong et al. proposed a framework that utilizes the structure feature learning model to capture the relations among the permission requests and its textual descriptions and functionalities. The work provides insights for users to be aware of potential risks of permission requests.

Defining the privacy setting in the right manner is the most important and cumbersome task for a naive user. Hence Anthonysamy et al. [4] have proposed CPM framework that helps the users to gain control over their data shared with TPAs by utilizing the social construct of friends to identify the best configurations and make use of the same.

Cheng et al. [7] have proposed a framework based on access control mechanism, wherein the applications are split into an internal and external component. Allowing the internal components to access private information but restricting the access to external parties. Implementing such a solution may help to limit the optional attributes, whereas limiting the required attributes leads to the user being denied to access the service. Another framework based on PBAC (Permission-Based Access Control) proposed by [16] Tomy et al., has been designed to give users complete control over their data and to decide on the information disclosure with third party applications. The work aims to provide the necessary awareness for the user in understanding the privacy risk involved in sharing sensitive data.

An approach similar to ours proposed by Egele et al. [8] have designed a browser plugin to intercept the data flow between OSN and TPA and through control mechanisms users can protect their profile data from malicious applications. The control mechanism of restricting the data once again stops the user from accessing the service. Whereas our approach does not limit the attribute, rather it grants access to the required attributes by generalizing them.

4   Conclusion

Preserving privacy in OSN is a challenging task. The very business model of OSN is based on analyzing the data shared by the user and using it as a monetizing fuel to run the business. The data is not only used within OSN but also the complexity increases by sharing it with third party application services to provide an extended service set to the users. Once the data is passed to TPA, the user's control over the data is lost, and they could even give the data to advertising agencies or data aggregating companies. The user's privacy threat spectrum widens with the inclusion of TPA. In this paper, we have collected information about the required attributes requested by the application. We have computed the privacy level of individual users and sensitivity of attributes to define the privacy parameter $\in$. Our proposed system Privacy Preserving Interceptor (PPI), intercepts the user's request and forwards the perturbed data to TPA. PPI perturbs the sensitive attribute by adding Laplacian noise or by replacing with random values. In our future extension, we wish to model the user's privacy level to include other features like the age of the user, gender, geographic location, cultural background. The attribute sharing is contextual. So, the TPA's trustworthiness could also be included as a parameter to decide the degree of disclosure. Improvements to perturbation techniques are being explored for certain attributes in which rather than replacing with random values, we are considering replacement with similar values to provide better utility.

References

1. Facebook platform. http://developers.facebook.com/
2. Teen cyberbullying (2016). http://www.deccanchronicle.com/lifestyle/health-and-wellbeing/230816/teen-cyberbullying-more-common-among-friends-dating-partners.html. Accessed 7 Aug 2016
3. Aldhafferi, N., Watson, C., Sajeev, A.: Personal information privacy settings of online social networks and their suitability for mobile internet devices. arXiv preprint arXiv:1305.2770 (2013)
4. Anthonysamy, P., Rashid, A., Walkerdine, J., Greenwood, P., Larkou, G.: Collaborative privacy management for third-party applications in online social networks. In: Proceedings of the 1st Workshop on Privacy and Security in Online Social Media, p. 5. ACM (2012)

5. Bilge, L., Strufe, T., Balzarotti, D., Kirda, E.: All your contacts are belong to us: automated identity theft attacks on social networks. In: Proceedings of the 18th International Conference on World Wide Web, pp. 551–560. ACM (2009)

6. Chaabane, A., Ding, Y., Dey, R., Kaafar, M.A., Ross, K.W.: A closer look at third-party OSN applications: are they leaking your personal information? In: Passive and Active Measurement, pp. 235–246. Springer, Cham (2014)

7. Cheng, Y., Park, J., Sandhu, R.: Preserving user privacy from third-party applications in online social networks. In: Proceedings of the 22nd International Conference on World Wide Web Companion, pp. 723–728. International World Wide Web Conferences Steering Committee (2013)

8. Egele, M., Moser, A., Kruegel, C., Kirda, E.: Pox: protecting users from malicious facebook applications. Comput. Commun. **35**(12), 1507–1515 (2012)

9. Mills, E.: Facebook suspends app. that permitted peephole (2008). http://news. cnet.com/8301-10784_3-9977762-7.htm. Accessed 10 May 2016

10. Jernigan, C., Mistree, B.F.: Gaydar: Facebook friendships expose sexual orientation. First Monday **14**(10) (2009)

11. Kelly: identity at risk on Facebook (2008). http://news.bbc.co.uk/2/hi/programmes/click_online/7375772.stm. Accessed 19 June 2015

12. Kong, D., Jin, H.: Towards permission request prediction on mobile apps via structure feature learning. In: Proceedings of SIAM International Conference on Data Mining (SDM 2015). SIAM (2015)

13. Kosinski, M., Stillwell, D., Graepel, T.: Private traits and attributes are predictable from digital records of human behavior. Proc. Nat. Acad. Sci. **110**(15), 5802–5805 (2013)

14. MadhumithaMurgia: cyber bullying (2016). http://www.telegraph.co.uk/technology/2016/06/19/facebook-leads-the-way-in-online-compassion-but-others-need-to-f/. Accessed 7 Aug 2016

15. Selkie, E.M., Fales, J.L., Moreno, M.A.: Cyberbullying prevalence among us middle and high school-aged adolescents: a systematic review and quality assessment. J. Adolesc. Health **58**(2), 125–133 (2016)

16. Tomy, S., Pardede, E., Taniar, D., Pardede, E.: Controlling privacy disclosure of third party applications in online social networks. Int. J. Web Inf. Syst. **12**(2) (2016)

17. Ybarra, M.L., Mitchell, K.J.: How risky are social networking sites? A comparison of places online where youth sexual solicitation and harassment occurs. Pediatrics **121**(2), e350–e357 (2008)

Multimodal Sentiment Analysis Using Deep Neural Networks

Harika Abburi[1]([✉]), Rajendra Prasath[2], Manish Shrivastava[1],
and Suryakanth V. Gangashetty[1]

[1] Langauage Technology Research Center,
International Institute of Information Technology Hyderabad, Hyderabad, India
harika.abburi@research.iiit.ac.in, {m.shrivastava,svg}@iiit.ac.in
[2] NTNU, Trondheim, Norway
drrprasath@gmail.com

Abstract. Due to increase of online product reviews posted daily through various modalities such as video, audio and text, sentimental analysis has gained huge attention. Recent developments in web technologies have also enabled the increase of web content in Hindi. In this paper, an approach to detect the sentiment of an online Hindi product reviews based on its multi-modality natures (audio and text) is presented. For each audio input, Mel Frequency Cepstral Coefficients (MFCC) features are extracted. These features are used to develop a sentiment models using Gaussian Mixture Models (GMM) and Deep Neural Network (DNN) classifiers. From results, it is observed that DNN classifier gives better results compare to GMM. Further textual features are extracted from the transcript of the audio input by using Doc2vec vectors. Support Vector Machine (SVM) classifier is used to develop a sentiment model using these textual features. From experimental results it is observed that combining both the audio and text features results in improvement in the performance for detecting the sentiment of an online product reviews.

Keywords: Multimodal sentiment analysis · MFCC · Doc2Vec · GMM · SVM · Deep neural networks

1 Introduction

Based on the opinion, Sentiment analysis classifies data into positive, negative and neutral categories. As of now most of the work on sentiment analysis is done on textual data. With increase in social media, people started sharing the information in the form of video, audio along with text. So all kinds of data are required for better sentiment classification. For any kind of approach like audio and text, sentiment can be extracted using sentiment classification techniques like lexicon based approach and machine learning approach [8].

R. Prasath—A part of this was carried out when the author was in Indian Institute of Information Technology (IIIT) Sricity, India.

© Springer International Publishing AG 2017
R. Prasath and A. Gelbukh (Eds.): MIKE 2016, LNAI 10089, pp. 58–65, 2017.
DOI: 10.1007/978-3-319-58130-9_6

Audio sentiment detection system is developed on the lines of Maximum Entropy modeling and Part Of Speech tagging. Transcripts from audio streams are obtained using Automatic Speech Recognition (ASR) [3]. In [4] rather than using ASR, Key Word Spotting (KWS) is used to extract the sentiment. Experiments have shown that the presented approach outperforms the traditional ASR. Authors of [6] have worked on determining if prosodic features can be used to build the sentiment classifier. Speech data which is there in audio file are generally extracted from the vocal track, excitation and prosody. Audio features like pitch, intensity and loudness are extracted using OpenEAR software and Support Vector Machine (SVM) classifier is built to detect the sentiment [12]. The audio features are automatically extracted from the audio track of each video clip using OpenEAR software and Hidden Markov Models (HMM) classifier is built to detect the sentiment [9]. In this paper Mel Frequency Cepstral Coefficients (MFCC) features are extracted and tested using DNN and GMM classifier.

Movie review mining using machine learning and semantic orientation is implemented [1]. In semantic orientation approach, bad and good associations account for negative and positive and based on the document features we can classify whether an input review belongs to a negative or positive class. Machine learning techniques are used to investigate the effectiveness of classification of documents by overall sentiment [10]. A variety of features like unigrams, bigrams and combination of both were employed, but the best results came from unigrams run through an SVM. Sentiment analyzer is developed to find out all the references on the subject and the sentiment polarity of each reference from online data documents [17]. To develop the sentiment analyzer, sentiment lexicon and the sentiment pattern database is used for extraction and association purposes. They classify expressions about specific items and use manually developed patterns to classify polarity. These patterns are high-quality, yielding a quite high precision, but very low recall. Sentiment is extracted using the opinion words like a combination of the adjectives along with the verbs and adverbs in the tweets [5]. They preprocess the tweets and add weightage according to the number of exclamation marks and the adjectives, verbs and adverbs are tagged in each tweet. Adjectives and negative words are taken into account to calculate the polarity of the whole phrase. The corpus-based method was used to find the semantic orientation of adjectives and the dictionary-based method to find the semantic orientation of verbs and adverbs. Two different Naive Bayes classifiers which make use of polarity lexicon to classif as positive and negative are used to detect the polarity of English tweets [2]. These classifiers are treated as the baseline. Features like lemmas, multiword, polarity lexicon and valence shifters are used. The training data set of tweets is obtained from SemEval 2014 and additional annotated tweets from external sources. Experiments show that performance is best when binary strategy is used with multiword and valence shifters. Approach to analyze the sentiment of short Chinese texts is presented in [15]. By using word2vec tool, sentiment dictionaries from NTU and HowNet are extended. Then the feature weight of the words are enhanced including the words that appear in the sentiment dictionary and the words next to the sentiment words. The model is implemented using SVM classifier.

The joint use of multiple modalities such as video, audio and text is explored for developing a sentimental model. Both feature level and decision level fusion methods are used to merge effective information extracted from multiple modalities. An improvement over classification by grouping over different modalities is reported in [14]. Multimodal sentiment analysis approach is an intelligent opinion mining system for identifying and understanding sentiment present in the reviews. In order to extract the sentiment they used audio and video signals and hence overcome the drawbacks of traditional sentiment analysis system [16]. In [7], the authors collected the English dataset from YouTube and expotv. The feature basis is formed by using text, video and audio features. Based on the textual movie review corpus, different levels of domain-dependence are considered such as in-domain analysis and cross-domain analysis. This shows that cross-corpus training works sufficiently well. Authors of [11] introduce database consisting of Spanish videos. They explored the combination of three modalities such as text, speech and video features on classification. They even explore the same work in English videos. From the results it is observed that the joint use of three modalities bring significant improvement. To determine the sentiment polarity present in the input [13] extracted the features from three modalities. The convolution neural network is used to extract the text features showed significant improvement in detecting the sentiment from a review.

In this work, a method to combine both the text and audio features is explored to detect the sentiment from the online product reviews. As of now, less research is done on the multimodal sentiment analysis of online reviews in Hindi language. Our proposed system is implemented in Hindi database. From the literature, it is observed that sentiment is detected from the input by extracting several features using OPENEAR/OPENSMILE tool and build a system using the SVM classifier. Instead of taking all the features, in this paper MFCC 13 dimension feature vector is extracted from each audio input and build sentiment analysis system using GMM and DNN classifiers. In order to detect the sentiment of a text input, textual features which are computed by Doc2Vec vectors are used to build the SVM classifier.

The rest of the paper is organized as follows: Hindi product reviews database used in this paper is discussed in Sect. 2. Sentiment analysis using audio features is discussed in Sect. 3. Sentiment analysis using text features is discussed in Sect. 4. Multimodal sentiment analysis and experimental results of proposed method for detecting the sentiment of a Hindi data is discussed in Sect. 5. Finally, Sect. 6 concludes the paper with a mention on the future scope of the present work.

2   Hindi Database

The database used in our studies is collected from YouTube, which is a publicly available source. The dataset includes reviews of phones, lotions and shampoos.

The database has some degree of generality as a variety of product reviews are used within the broad domain of product reviews. The two basic sentiments presented in the database are: Positive and Negative. The average length of each input is thirty seconds and average number of words in each input is around 40. A total of 110 product reviews are collected, among them 100 inputs are taken based on inter-annotator agreement. Transcription and sentiment annotations were manually performed for text based sentiment classification. Both the modalities such as audio and text are provided for annotators to figure out the exact opinion of the input. Then based on inter-annotator agreement, 50 positive and 50 negative inputs are selected. Among them 80% are used for training and remaining 20% are used for testing.

3 Sentiment Analysis Using Audio Features

The process of developing the sentiment model and extracting the audio features from the input is described in this section. The features which are extracted are used to build a classifier of positive or negative sentiment. Each input is in the form of .wav format, in 16 bit, 16000 Hz sampling frequency and a mono channel. MFCC features are extracted from each input and sentiment model is developed using Gaussian Mixture Models (GMM) and Deep Neural Network classifiers. Block diagram of sentiment analysis using Deep Neural Network is shown in the Fig. 1.

A deep neural network (DNN) is a neural network with multiple hidden layers of nodes between the input and output layers. These hidden layers do feature identification and processing in a series of stages. The successive layers can learn higher level features. DNN performance depends on training data. The more the training data the more accurate it was. Each DNN is trained for 30 epochs with different number of layers and different number of nodes in each layer. In our work up to four deep layers are explored. All the DNNs are trained with ADAM method which is hyper parameter learning algorithm. MFCC features considered in this study are 13-Dimension, 65-Dimension and 130-Dimension. 10 frames of 13 dimension MFCC frames are concatenated to get 130 dimension and 5 frames of 13 dimension MFCC frames are concatenated to get 65 dimension. Frames are concatenated here because each frame will not carry the sentiment. So, experiments are even done in combination of frames, which results in better performance. Based on the input dimension, the input layer nodes can be 13, 65 and 130 which are linear. The output layer is of softmax layer with 2 nodes because the number of classes in the database are 2. During testing the node which gives maximum score is assigned as the claimed class. For testing only 5 s of data is taken for each input.

MFCC features which are extracted from each input is given as input to the GMM. GMM is tested with different number of test cases and different number of mixtures like 16, 32 and 64. Here also for testing 5 s of data is taken.

From the Table 1 it is observed that DNN with 65-Dimension feature vector has performed better when compared to 130 dimensions because by using 130

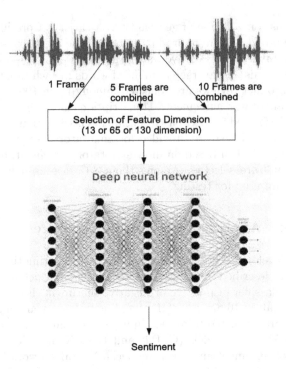

Fig. 1. Block diagram of sentiment analysis using deep neural network

Table 1. Performance of sentiment analysis using deep neural network

DNN	Two layers (%)	Three layers (%)
13-Dimension	41.66	58.3
65-Dimension	58.3	**75.0**
130-Dimension	58.3	66.7

Table 2. Performance of sentiment analysis using classifiers

Classifier	Accuracy (%)
GMM (64)	58.3
DNN (three layer)	**75.0**

dimension feature vector, features are not sufficient to train the DNN. It is also observed that performance is more with three layers. The fourth layer is also explored but we are getting same accuracy as in the third layer. From the Table 2 it is observed that DNN with three hidden layers outperforms the GMM with 64 mixtures by 17%.

4   Sentiment Analysis Using Text Features

The process of developing the sentiment model and extracting the text features from the input is described in this section. These features are used to build a classifier of positive or negative sentiment. In a preprocessing step, each audio input is manually transcribed and sentiment annotations are also assigned manually. For better results, 300 dimension feature vector is generated from each text input.

Many machine learning algorithms require the input to be represented as a fixed-length feature vector. Word2Vec takes data from a corpus, and churns out vectors for each of those words. These vectors are interesting because similar words are placed nearer to each other in the vector space. Doc2Vec model represents not only words, but it is an unsupervised learning of continuous representations for larger text such as sentences, paragraphs and whole documents. In Doc2Vec architecture, the algorithms used are distributed memory and distributed bag of words. Distributed memory will randomly initialize paragraph vector for each document and predict next instance using context words and paragraph vectors. Context window slide across document, but paragraph vector is fixed. On the other hand distributed bag of words will only use paragraph vectors and not word vectors. It will take window of words in a paragraph and randomly sample which one to predict using paragraph vector (ignores word ordering). By combining both the algorithms Doc2Vec generates the vectors.

Doc2Vec will generate a single vector for each manually transcribed input, which represents the meaning of a document. To associate documents with labels this vector will be used as an input to a supervised machine learning algorithm. Sentiment analysis based on text can be viewed as a text classification task which can be handled by SVM. SVM classifier is trained with vectors generated from the doc2vec and by using corresponding sentiment tags as positive or negative. Given a test input, the trained models classify it as either positive or negative. From Table 3 it is observed that the rate of detecting the sentiment of Hindi language product reviews for text features is 63.64 %.

Table 3. Performance of sentiment analysis using text features

Classifier	Accuracy (%)
SVM	63.64

5   Multimodal Sentiment Analysis

Analyzing the audio data has advantage of voice modularity compared to the textual data. In textual data, exact sentiment of the input may not extract properly because it only has the information regarding the words and their dependencies. Instead, audio data contain multiple modalities like acoustic and

Table 4. Performance of multimodal sentiment analysis

Modality	Accuracy (%)
Audio	75
Text	63.4
Audio + text	**78.2**

linguistic streams. Both the modalities are hypothesized based on the highest average probability of the classifiers. From our experiments, it is observed that the simultaneous use of these two modalities help to create a better sentiment analysis model to detect whether the given test input is positive or negative sentiment.

From the Table 4 it is observed that by combining the two modalities such as text and audio rate of detecting the sentiment of a product review is improved.

6   Conclusion

In this paper, we proposed an approach to extract the sentiment of a given input using both audio and text information. MFCC features are extracted from audio and sentiment models are built using DNN and GMM. DNN is tested with different layers with different number of nodes, whereas GMM is tested with different mixture components with different test cases. For text, features generated using Doc2Vec are used to build the model using the SVM classifier. From our experiments, it is observed that DNN classifier with 65-dimension MFCC features has high accuracy of detecting the sentiment of an input compared to other dimensions and even DNN classifier outperformed GMM classifier by 17%. It is also observed that by combining both the modalities such as audio and text, rate of detecting the sentiment is significantly improved.

References

1. Chaovalit, P., Zhou, L.: Movie review mining: a comparison between supervised and unsupervised classification approaches. In: Proceedings of IEEE 38th Hawaii International Conference on System Sciences, Big Island, Hawaii, pp. 1–9 (2005)
2. Gamallo, P., Garcia, M.: Citius: a naive-bayes strategy for sentiment analysis on english tweets. In: Proceedings of the 8th International Workshop on Semantic Evaluation (SemEval 2014), pp. 171–175, August 2014
3. Kaushik, L., Sangwan, A., Hansen, J.H.L.: Sentiment extraction from natural audio streams. In: Proceedings of IEEE International Conference on Acoustics Speech and Signal Processing (ICASSP), pp. 8485–8489 (2013)
4. Kaushik, L., Sangwan, A., Hansen, J.H.: Automatic audio sentiment extraction using keyword spotting. In: Proceedings of Interspeech, pp. 2709–2713, September 2015
5. Kumar, A., Sebastian, T.M.: Sentiment analysis on twitter. IJCSI Int. J. Comput. Sci. **9**(4), 372–378 (2012)

6. Mairesse, F., Polifroni, J., Fabbrizio, G.D.: Can prosody inform sentiment analysis? experiments on short spoken reviews. In: Proceedings of IEEE International Confernce on Acoustics, Speech and Signal processing (ICASSP), pp. 5093–5096 (2012)

7. Wollmer, M., Felix, W., Knaup, T., Morency, L.P.: YouTube movie reviews: sentiment analysis in an audio-visual context. IEEE Intll. Syst. **28**(3), 46–53 (2013)

8. Medhat, W., Hassan, A., Korashy, H.: Sentiment analysis algorithms and applications: a survey. Ain Shams Eng. J. **5**, 1093–1113 (2014)

9. Morency, L.P., Mihalcea, R., Doshi, P.: Towards multimodal sentiment analysis: harvesting opinions from the web. In: Proceedings of the 13th International Conference on Multimodal Interfaces (ICMI2011), pp. 169–176, November 2011

10. Pang, B., Lee, L., Vaithyanathan, S.: Thumbs up?: sentiment classification using machine learning techniques. In: Proceedings of ACL-02 Conference on Empirical Methods in Natural Language Processing, vol. 10, pp. 79–86 (2002)

11. Perez-Rosas, V., Mihalcea, R., Morency, L.P.: Multimodal sentiment analysis of spanish online videos. IEEE Intll. Syst. **28**(3), 38–45 (2013)

12. Perez-Rosas, V., Mihalcea, R., Morency, L.P.: Utterance level multimodal sentiment analysis. In: Proceedings of ACL, pp. 973–982 (2013)

13. Poria, S., Cambria, E., Gelbukh, A.: Deep convolutional neural network textual features and multiple kernel learning for utterance-level multimodal sentiment analysis. In: Proceedings of EMNLP, pp. 2539–2544 (2015)

14. Poria, S., Cambria, E., Howard, N., Huang, G.B., Hussain, A.: Fusing audio, visual and textual clues for sentiment analysis from multimodal content. Neurocomputing **174**, 50–59 (2015)

15. Xing, L., Yuan, L., Qinglin, W., Yu, L.: An approach to sentiment analysis of short chinese texts based on SVMs. In: Proceedings of the 34th Chinese Control Conference, pp. 28–30. IEEE, July 2015

16. Yadav, S.K., Bhushan, M., Gupta, S.: Multimodal sentiment analysis: sentiment analysis using audiovisual format. In: Proceedings of IEEE 2nd International Conference on Computing for Sustainable Global Development (INDIACom), pp. 1415–1419 (2015)

17. Yi, J., Nasukawa, T., Bunescu, R., Niblack, W.: Sentiment analyzer: extracting sentiments about a given topic using natural language processing techniques. In: Proceedings of IEEE International Conference on Data Mining (ICDM) (2003)

A Study on Text-Independent Speaker Recognition Systems in Emotional Conditions Using Different Pattern Recognition Models

K.N.R.K. Raju Alluri[1(✉)], Sivanand Achanta[1], Rajendra Prasath[2],
Suryakanth V. Gangashetty[1], and Anil Kumar Vuppala[1]

[1] Speech and Vison Lab (LTRC),
International Institute of Information Technology Hyderabad,
Hyderabad, Andhra Pradesh, India
{raju.alluri,sivanand.a}@research.iiit.ac.in,
{svg,anil.vuppala}@iiit.ac.in
[2] NTNU, Trondheim, Norway
drrprasath@gmail.com

Abstract. The present study focuses on the text-independent speaker recognition in emotional conditions. In this paper, both system and source features are considered to represent speaker specific information. At the model level, Gaussian Mixture Models (GMMs), Gaussian Mixture Model-Universal Background Model (GMM-UBM) and Deep Neural Networks (DNN) are explored. The experiments are performed using 3 emotional databases, i.e. German emotional speech database (EMO-DB), IITKGP-SESC: Hindi and IITKGP-SESC: Telugu databases. The emotions considered in the present study are neutral, anger, happy and sad. The results show that, the performance of a speaker recognition system trained with clean speech is degrading while testing with emotional data irrespective of feature used or model used to build the system. The best results are obtained for the score level fusion of system and source features based systems when speakers are modeled with DNNs.

Keywords: Speaker recognition · Emotion · System features · Source features · Gaussian Mixture Modeling · Universal Background Modeling · Deep Neural Networks

1 Introduction

Every individual has a unique identity in his voice. This property of speech gives us a scope to recognize a person from his voice without looking at him, i.e. we can recognize a friend over the phone. The process of recognizing a person from his own voice is known as Speaker Recognition (SR) [1]. SR task can be subdivided into speaker verification and speaker identification. Speaker verification

R. Prasath—A part of this was carried out when the author was in Indian Institute of Information Technology (IIIT) Sricity, India.

R. Prasath and A. Gelbukh (Eds.): MIKE 2016, LNAI 10089, pp. 66–73, 2017.
DOI: 10.1007/978-3-319-58130-9_7

is the task of determining whether the person is the claimed or not. Speaker identification is the task of determining who is speaking in the set of known speakers [19]. SR systems can be further classified into either text dependent or text independent based on the text to be spoken. In text dependent mode, both training and testing use same text, where as in text independent case there are no such restrictions [21]. In the present work text-independent speaker identification approach is considered. Applications of SR include access control, transaction authentication, law enforcement, speech data management etc.

Development of the SR system uses different features which are extracted from the speech signal. As speech signal is the result of exciting a time varying vocal tract system with time varying excitation [17]. The information regarding speaker will be there in both system and source features. Most of the SR systems are modeled using Mel Frequency Cepstral Coefficients (MFCCs) and Linear Prediction Cepstral Coefficients (LPCCs) which represent vocal tract characteristics [1,16]. In [2,15,18] features are extracted from Linear Prediction (LP) residual as it contains significant information about excitation source to model the SR systems. The features which capture excitatilon information are Real Cepstral Coefficients (RCCs). The state of the art SR systems such as i-vector [5], GMM-UBM [20] and Gaussian Mixture Model-Support Vector Machine (GMM-SVM) [4] have achieved significant performance. In [7] DNNs are used for acoustic modeling in speech recognition. Speaker specific characteristics using DNNs are explored in [23]. The success of DNNs in speech recognition opened the window for researchers to use DNNs for other speech related tasks like speaker recognition and language identification [8,12,14,22].

The performance of SR depends on the features, models and the conditions in which training and testing are done. The inconsistency in training and testing models causes the performance degradation in SR tasks. The factors that cause performance degradation include background noise, channel effect and speaker health condition [25]. Along with these factors, emotional state of the speaker is one of the reason which causes degradation in SR performance. In the literature, it is reported that the performance of SR systems was degraded in emotional conditions [6,11]. In order to improve SR system performance, emotion state conversion incorporated into the models was reported in [13] and The effect of adding emotional data to the neutral data in training is analyzed in [24,26].

In the previous studies, researchers used MFCCs as features and modeling techniques are GMM, GMM-UBM, SVM for SR in emotional conditions. Our contributions in this study are firstly, DNNS are explored for SR in emotional conditions. Second contribution in the present study is to get benefit from the complimentary nature of the features extracted from the system and source, we have fused the scores of these two systems which improved the performance of SR system.

The rest of the paper is organized as follows. Databases used in this study are described in Sect. 2. SR system is described in Sect. 3. Experimental setup is explained in Sect. 4. Section 5 presents SR in emotional conditions followed by the results of the present work. Conclusion and future score are addressed in Sect. 6.

2   Emotional Databases

The experiments are conducted using three different emotion databases from three different languages. The databases used in this study are IITKGP-SESC: Hindi database [9], IITKGP-SESC: Telugu database [10] and Emo-DB: Berlin database [3]. Each database contains 10 professional speakers (5 male and 5 female) data in several emotions. For recording the emotions in IITKGP-SESC:Hindi and Telugu databases artists are provided with 15 text prompts which are neutral in meaning. The artists have to speak these 15 utterances in 8 (anger, disgust, fear, happy, sad, sarcastic, surprise and neutral) basic emotions in one session. There are total 10 sessions for each speaker in all emotions. In case of EMO-DB speakers were asked to speak 10 sentences in 7 (angry, happy, neutral, sad, fear, disgust and boredom) different emotions, in one or more sessions. In this study, we have considered 4 emotions (neutral, happy, anger and sad). All speech recordings are done at 16 Khz sampling rate and a bit rate of 16 bits/samples. For each speaker we have considered 10 utterances in each emotion.

3   Speaker Recognition Description

SR system will have two phases in its development. The first one is training/enrolling phase and the other is testing phase.

Enrollment Phase. Enrollment phase contains two basic building blocks. The first one is front-end processing and the second one is modeling. In front end processing the speech signal is passed through a speech activity detector to remove the silence and non-speech regions from signal later the features are extracted from small segments of signal to ensure the quasi stationary nature of the speech signal. In the modeling the features of each speaker are modeled using a pattern classification algorithm. At the end of the enrollment phase, each speaker will have one model (Fig. 1).

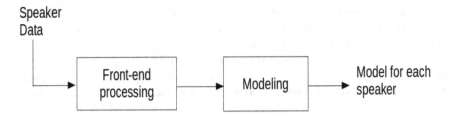

Fig. 1. Enrollment phase of speaker recognition

Testing Phase. In the testing phase the test speaker speech signal is passed through the front-end processing to remove non speech regions and the features are extracted in the same way as in enrollment phase. These features are considered to get scores from each model present in enrollment phase, the model which gives maximum score will be the claimed speaker (Fig. 2).

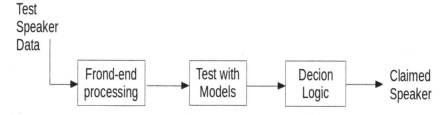

Fig. 2. Testing phase of speaker recognition

4   Experimental setup

4.1   Baseline SR system

In the present study, we considered 30 speakers neutral data from the standard emotional databases. Each speaker has 10 utterances each of approximately 3 s duration. We consider 8 utterances for training and the remaining 2 utterances for testing. That means for training we have considered 24 s data and testing 6 s data. We have considered different combinations for this data and the average values are reported as the final results in Table 1. The features used in this study are MFCCs and RCCs. Pattern classification algorithms like Gaussian Mixture Models (GMMs), Gaussian Mixture Model-Universal Background Model (GMM-UBM) and Deep Neural Networks (DNN) are used as modeling techniques. The results show that performance of SR task has achieved maximum accuracy when trained and tested with neutral utterances. The details of these experimental setups are described in the following subsections.

Table 1. Performance (%) of SR system when trained and tested with neutral utterances

Modeling technique	MFCC	RCC
GMM	98.33	96.66
GMM-UBM	100	98.33
DNN	98.33	95

SR System Using GMM. Energy based threshold is applied as a speech activity detector in this study. GMMs with 8,16 and 32 mixture components are build for each speaker using 13-Dimensional MFCC features extracted from the speech signal of 20 ms frame size with a 5 ms frame shift. During Testing log likelihoods are computed for each model and the one which give the max score is assigned as the claimed speaker. The same procedure repeated for RCC also. The best results occurred for 13-Dimensional features with 16 Mixture components are reported in Table 1.

SR System Using GMM-UBM. In this case also we employed energy based thresholding as SAD. UBM is trained on 632 speakers data from TIMIT using 256, 512 and 1024 mixtures, then by using Maximum A Posteriori (MAP) we adapted GMM for each speaker. In this case also we considered both MFCCs and RCCs. During Testing log likelihoods are computed for each model and is removed from UBM score and the one which gives the max score is assigned as the claimed speaker. 512 Mixture UBM is performing better than the others so the results reported accordingly in Table 1.

SR System Using DNN. For DNN also we considered both MFCCs and RCCs the SAD is also same as mentioned previously. Each DNN was trained for 50 epochs with different number of layers and different number of nodes in each layer. All the DNNs are trained with hyper parameter learning algorithms like SGD-CM and ADAM method. As the MFCC features considered in this study are of 13-Dimension the input layer contains 13 nodes and linear. The output layer is of softmax layer with 30 nodes because the number of speakers in database are 30. During Testing the node which gives maximum score is assigned as the claimed speaker. The results reported in Table 1 are with 4 hidden layers.

5   Speaker Recognition in Emotional Conditions

In this section three scenarios are considered to study the influence of emotion on features and modeling techniques used to build the SR system. In all the scenarios speaker models are trained with neutral speech and tested with 3 (anger, happy and sad) emotional speech utterances. The experimental setup is same as that of baseline system.

Table 2. Performance (%) of SR system in emotional conditions using MFCC

Emotion	GMM	GMM-UBM	DNN
Anger	21.66	28.33	**38.33**
Happy	38.33	38.33	**48.33**
Sad	38.33	36.66	**63.33**

In the first scenario the speaker specific information is represented using MFCCs and GMMs, GMM-UBM and DNNs are used as modeling techniques. The experimental results are reported in Table 2. From the experimental results it can be observed that the performance degradation was observed due to emotion present in the speech signal. The average performance degradation in case of GMMs, GMM-UBM and DNNs are 66%, 64% and 48% respectively. From the results it can also be observed that DNNs are performing better than the other modeling techniques by a margin of around 15%.

Table 3. Performance (%) of SR system in emotional conditions using RCC

Emotion	GMM	GMM-UBM	DNN
Anger	16.66	25	**43.33**
Happy	31.66	36.66	**46.66**
Sad	30	36.66	**50**

In the second scenario the features used are RCCs and modeling techniques used are GMM, GMM-UBM and DNNs. The experimental results are reported in Table 3. In this scenario also the performance degradation was observed. There is a degradation in the performance of the SR system from 96.66% to 26% in GMMs, 98.33 to 32.77% in GMM-UBM and 95 to 46.66 in DNNs. In this scenario also DNNs are giving better results than the GMM and GMM-UBM. From the two scenarios, it is clear that the SR performance is affected by the emotion present in the speech. DNNs are modeling the speaker specific characteristics more robustly than the GMM and GMM-UBM.

As MFCCs represents vocal tract information present in speech and RCCs represent excitation source information present in speech they are complimentary in nature. As compliment systems are there to study SR task. A system fusion is performed on the MFCC base system and RCC based system at score level. The experimental results are given in Table 4. From the results it can be observed that fusion of MFCC and RCC based systems are improving the accuracy of the SR system by 7% in GMM, 9% in GMM-UBM and 5% in case of DNNs.

Table 4. Performance (%) of SR system in emotional conditions

Emotion	GMM			GMM-UBM			DNN		
	MFCC	RCC	MFCC+RCC	MFCC	RCC	MFCC+RCC	MFCC	RCC	MFCC+RCC
Anger	21.66	16.66	**30**	28.33	25	**38.33**	38.33	43.33	**45**
Happy	38.33	31.66	**43.33**	38.33	36.66	**45**	48.33	46.66	**50**
Sad	38.33	30	**38.33**	36.66	36.66	**41.66**	63.33	50	**65**

6 Summary and Future Scope

In this paper, both system and source features are used to represent speaker specific information. Pattern classification algorithms like GMM, GMM-UBM and DNNs are used as the modeling techniques. The experiments are conducted using 30 (15 male and 15 male) speakers data taken from 3 (EMO-DB,IITKGP-SESC: Hindi and Telugu) different emotional databases. The speaker models are trained with neutral speech and tested with 4 (neutral, anger, happy and sad) utterances. Experimental results show that when trained and tested with neutral utterances there is a significant accuracy irrespective of features and modeling techniques used. Where as when the speaker models are trained with neutral utterances and tested with emotional utterances there is a degradation in performance. The performance degradation was less in DNNs compared to other modeling techniques. Further the performance of SR system is improved by fusing the scores of MFCC and RCC based systems. In our future work different architectures of DNNs will be explored to capture more speaker specific characteristics.

Acknowledgements. The first author would like to thank Department of Electronics and Information Technology, Ministry of Communication & IT, Govt of India for granting PhD Fellowship under Visvesvaraya PhD Scheme.

References

1. Atal, B.S.: Automatic recognition of speakers from their voices. Proc. IEEE **64**(4), 460–475 (1976)
2. Atal, B.S.: Automatic speaker recognition based on pitch contours. J. Acoust. Soc. Am. **52**(6B), 1687–1697 (1972)
3. Burkhardt, F., Paeschke, A., Rolfes, M., Sendlmeier, W.F., Weiss, B.: A database of german emotional speech. In: Proceedings of the INTERSPEECH, vol. 5, pp. 1517–1520 (2005)
4. Campbell, W.M., Sturim, D.E., Reynolds, D.A.: Support vector machines using gmm supervectors for speaker verification. IEEESignal Process. Lett. **13**(5), 308–311 (2006)
5. Dehak, N., Kenny, P.J., Dehak, R., Dumouchel, P., Ouellet, P.: Front-end factor analysis for speaker verification. IEEE Trans. Audio Speech Lang. Process. **19**(4), 788–798 (2011)
6. Ghiurcau, M.V., Rusu, C., Astola, J.: Speaker recognition in an emotional environment. In: Proceedings of the Signal Processing and Applied Mathematics for Electronics and Communications, pp. 81–84 (2011)
7. Hinton, G., Deng, L., Yu, D., Dahl, G.E., Mohamed, A., Jaitly, N., Senior, A., Vanhoucke, V., Nguyen, P., Sainath, T.N., et al.: Deep neural networks for acoustic modeling in speech recognition: The shared views of four research groups. IEEE Signal Process. Mag. **29**(6), 82–97 (2012)
8. Lakshmi, H.R., Achanta, S., Bhavya, P.V., Gangashetty, S.V.: An investigation of end-to-end speaker recognition using deep neural networks. Int. J. Eng. Res. Electron. Commun. Eng. **3**(1), 42–47 (2016)

9. Koolagudi, S.G., Krothapalli, R.S.: Two stage emotion recognition based on speaking rate. Int. J. Speech Technol. **14**(1), 35–48 (2011)
10. Koolagudi, S.G., Maity, S., Kumar, V.A., Chakrabarti, S., Rao, K.S.: IITKGP-SESC: Speech database for emotion analysis. In: Ranka, S., Aluru, S., Buyya, R., Chung, Y.-C., Dua, S., Grama, A., Gupta, S.K.S., Kumar, R., Phoha, V.V. (eds.) IC3 2009. CCIS, vol. 40, pp. 485–492. Springer, Heidelberg (2009). doi:10.1007/978-3-642-03547-0_46
11. Koolagudi, S.G., Sharma, K., Sreenivasa Rao, K.: Speaker recognition in emotional environment. In: Mathew, J., Patra, P., Pradhan, D.K., Kuttyamma, A.J. (eds.) ICECCS 2012. CCIS, vol. 305, pp. 117–124. Springer, Heidelberg (2012). doi:10.1007/978-3-642-32112-2_15
12. Mounika, K.V., Achanta, S., Lakshmi, H.R., Suryakanth, V.G., Vuppala, A.K.: An investigation of deep neural network architectures for language recognition in Indian languages. In: Proceedings of the INTERSPEECH, pp. 2930–2933 (2016)
13. Li, D., Yang, Y., Wu, Z., Wu, T.: Emotion-state conversion for speaker recognition. In: Tao, J., Tan, T., Picard, R.W. (eds.) ACII 2005. LNCS, vol. 3784, pp. 403–410. Springer, Heidelberg (2005). doi:10.1007/11573548_52
14. Lopez-Moreno, I., Gonzalez-Dominguez, J., Plchot, O., Martinez, D., Gonzalez-Rodriguez, J., Moreno, P.: Automatic language identification using deep neural networks. In: Proceedings of the ICASSP, pp. 5337–5341 (2014)
15. Makhoul, J.: Linear prediction: A tutorial review. Proc. IEEE **63**(4), 561–580 (1975)
16. Oshaughnessy, D.: Speaker recognition. IEEE ASSP Mag. **3**, 4–17 (1986)
17. O'shaughnessy, D.: Speech Communication: Human and Machine. Universities Press, India (1987)
18. Prasanna, S.M., Gupta, C.S., Yegnanarayana, B.: Extraction of speaker-specific excitation information from linear prediction residual of speech. Speech Commun. **48**(10), 1243–1261 (2006)
19. Reynolds, D.: An overview of automatic speaker recognition. In: Proceedings of the ICASSP, pp. 4072–4075 (2002)
20. Reynolds, D.A., Quatieri, T.F., Dunn, R.B.: Speaker verification using adapted gaussian mixture models. Digit. Signal Process. **10**(1), 19–41 (2000)
21. Reynolds, D.A., Rose, R.C.: Robust text-independent speaker identification using gaussian mixture speaker models. IEEE Trans. Speech Audio Process. **3**(1), 72–83 (1995)
22. Richardson, F., Reynolds, D., Dehak, N.: Deep neural network approaches to speaker and language recognition. IEEE Signal Process. Lett. **22**(10), 1671–1675 (2015)
23. Salman, A., Chen, K.: Exploring speaker-specific characteristics with deep learning. In: Proceedings of the IJCNN, pp. 103–110. IEEE (2011)
24. Scherer, K.R., Johnstone, T., Klasmeyer, G., Bänziger, T.: Can automatic speaker verification be improved by training the algorithms on emotional speech? In: Proceedings of the INTERSPEECH, pp. 807–810 (2000)
25. Wegmuller, M., von der Weid, J.P., Oberson, P., Gisin, N.: Study on speaker verification on emotional speech. In: Proceedings of the INTERSPEECH (2006)
26. Wu, T., Yang, Y., Wu, Z.: Improving speaker recognition by training on emotion-added models. In: Tao, J., Tan, T., Picard, R.W. (eds.) ACII 2005. LNCS, vol. 3784, pp. 382–389. Springer, Heidelberg (2005). doi:10.1007/11573548_49

A Study on Vowel Region Detection from a Continuous Speech

Ramakrishna Thirumuru$^{(\boxtimes)}$, Harikrishna Vydana,
Suryakanth V. Gangashetty, and Anil Kumar Vuppala

Speech and Vision Lab (LTRC), International Institute
of Information Technology Hyderabad, Hyderabad, India
{ramakrishna.thirumuru,hari.vydana}@research.iiit.ac.in,
{svg,anil.vuppala}@iiit.ac.in

Abstract. Vowels exhibit high sonority and loudness due to varying strength of impulse-like excitations. Acoustic events such as vowel onset point (VOP) and vowel end point (VEP) can be used to detect precise vowel regions in a speech signal. In this paper, a technique is proposed to detect vowel regions based on these acoustic parameters in a continuous speech signal. Vowels possess significant energy content in the low frequency bands of speech. The initial phase of this method consists of speech signal processing using zero frequency filtering technique. Zero frequency filtered signal predominantly contains low frequency content of the speech signal as it is filtered around 0 Hz. This process is followed by the extraction of dominant spectral peaks from the magnitude spectrum around glottal closure regions of the speech signal. The vowel onset points and vowel end points are obtained by convolving spectrum of zero frequency filtered signal with first order Gaussian differentiator. The performance of the proposed vowel region detection method is compared with the existing state of art methods on TIMIT database. It is reported that this method produced relatively significant improvement in vowel region detection in clean and noisy environments.

Keywords: Vowel Onset Point (VOP) · Vowel End Point (VEP) · Zero frequency filtering · Magnitude spectrum · First order Gaussian differentiator

1 Introduction

Vowel region detection is a task of identifying a vowel occurrences in a continuous speech with precise boundary markings. VOP and VEP can be considered as vowel boundaries in a continuous speech. The importance of vowel boundaries for speech analysis can be felt in many applications. The detection of vowels in a speech is an important step in automatic speech recognition, smart audio filtering, recognition of CV units in Indian languages, determining the duration of vowels, speech rate manipulation and multimedia synchronization [1,4,6,7]. The knowledge of phonetic classes can be integrated into the statistical based ASR

© Springer International Publishing AG 2017
R. Prasath and A. Gelbukh (Eds.): MIKE 2016, LNAI 10089, pp. 74–82, 2017.
DOI: 10.1007/978-3-319-58130-9_8

systems to improve the overall performance of the system. This mechanism can provide a platform to automatic speech recognition systems to utilize segment based approach instead of dominant frame based HMM models [2,8].

Acoustic-phonetic feature based landmark detection has received considerable attention in recent years. In the literature, the problem of identifying vowel regions under the context of landmark detection is studied through the extraction of distinctive features. Several vowel region detection methods have been proposed based on acoustic features in the literature and we refer to some of them here. Vowel onset point detection methods are based on rising slope of spectral amplitude in the magnitude spectrum of the speech signal. A method is proposed by combining evidence from the excitation source, spectral peaks and modulation spectrum for the detection of vowel onset point [5]. Improved vowel onset point detection for vowel region extraction is proposed based on spectral energy present in glottal closure regions of speech signal [9,10]. In another attempt, a method was proposed to detect vowel end points along with vowel onset points based on falling slope of spectral energy [11]. The region between vowel onset point and vowel offset point is considered as a vowel region. These methods report that most vowel onset and end points fall within 40 ms deviation.

In this paper, we present another method for the detection of speaker independent vowel regions using acoustic features. This approach is based on detection of landmarks such as vowel onset point and vowel end point with an assumption that vowels exhibit high sonority and loudness. A combined approach is proposed from previous works, using the features of zero frequency filtered signal for vowel region detection. These landmarks are detected based on spectral content intensity variation around glottal closure instants of zero frequency filtered speech signal. Zero frequency filtered speech signal around glottal closure regions is processed to bring out high information pertaining to the vowel regions.

This paper is organized as follows: Sect. 2 describes baseline methods for detecting vowel regions using vowel onset points and end points. The proposed method for vowel region detection is described in Sect. 3. The performance of the proposed method is studied using TIMIT database and, results are discussed in Sect. 4. Section 5 provides summary and conclusions of this work.

2   Baseline Methods for Vowel Region Detection Using Vowel Onset Point and Vowel End Point

In this Section, state of art methods for vowel detection using VOP and VEP in a continuous speech are discussed. These methods are based on previous works mentioned in Refs. [5,9]. The experimental results obtained for the proposed method are compared with these methods.

2.1   Vowel Region Detection by Combining the Evidences from Excitation Source, Spectral Peaks and Modulation Spectrum

The combined evidence for the boundary markings of a vowel region is derived from three parameters namely excitation source, spectral peaks and modulation spectrum. Therefore it is termed as COMB method in this paper. Hilbert envelope of linear prediction (LP) residual otherwise known as excitation source is estimated. The smoothened Hilbert envelope of the LP residual is obtained by convolving with Hamming window of 50 ms. This is enhanced using first order difference (FOD) to improve the detection of VOP and VEP. The enhanced signal is convolved with first order Gaussian difference (FOGD) operator, and convolved output is the evidence for VOP and VEP. These acoustic events are based on the nature of the gradient of the output signal.

A 256-point DFT of speech signal with 20 ms duration with 50% overlap produces amplitude spectrum. The sum of ten largest spectral peaks are selected from first 128 points and plotted as a function of a time, and this represent the energy of the spectral peaks. The VOP and VEP can be observed as significant change in complimentary manner in this time varying signal. The changes in VOP and VEP are enhanced in the spectral energy signal using FOD. The variations that represent VOP and VEP are extracted using FOGD operator.

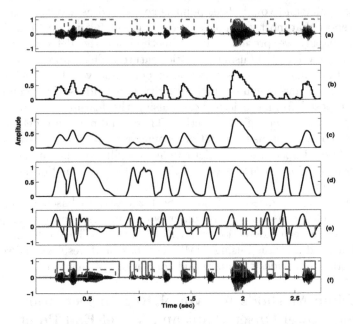

Fig. 1. Vowel region detection method using the COMB method. (a) speech signal with manually marked vowel regions. (b) combined evidence. (c) mean smoothed evidence contour. (d) enhanced evidence using FOD. (e) VOP and VEP marking for the speech signal. (f) predicted vowel regions.

The information in speech signal can also be represented by changes in the time domain envelope and this is referred as amplitude-modulation spectrum or simply modulation spectrum. The modulation spectrum of speech is dominated by the low-frequency components. The speech signal is analyzed using

18 trapezoidal critical band filters between 0 and 4 kHz. An amplitude envelope signal is computed using half wave rectification and low pass filtering on all bands. Amplitude envelope signals are down-sampled to 80 samples/s and normalized by the average envelope of that channel, measured over entire utterance. The modulations of the normalized envelope signals are analyzed by computing DFT over 250 ms with an overlap of 5% in order to capture dynamic properties of the signal. The 4–16 Hz components are added together across all critical bands to derive modulation spectrum energy. Thus obtained signal is enhanced and processed to obtain evidence for VOP and VEP using FOD and FOGD respectively. The combined method uses three independent and complementary evidences to derive a single combined evidence. This is obtained by adding three shreds of evidence sample by sample and the performance of detection of VOP and VEP is better than individual evidences. The peaks and valleys of combined evidence are marked as VOP and VEP respectively. Experimental result for the COMB method is shown in Fig. 1. Figure 1(a),(b),(c),(d) refers to a continuous speech utterance, combined evidence, mean smoothed evidence contour, enhanced evidence using first order Gaussian difference, VOP and VEP marking for the speech signal and prediction respectively. This method produced high false alarm due to spurious peaks.

2.2 Vowel Region Detection from VOP and VEP Using Spectral Energy Around Glottal Closure Instants

The vocal tract is completely isolated from trachea and lungs during glottal closure phase. Spectrum estimation during glottal closure phase will be more accurate as true vocal tract resonances are present in during this period. In this method, spectral energy at the glottal closure instants is used as an evidence to detect VOP and VEP in a continuous speech. Firstly, the glottal closure instants are extracted using zero frequency filtering. Around the glottal closure instants, formants are computed for 30% of speech samples using group delay function and it is termed as FGCI method. The spectral energy of speech signal is derived from the first three formants and the same is plotted as a function of time. This contour is smoothed using mean smoothing window of 50 ms and enhanced using FOD operator. Significant changes in the spectral characteristic present in the enhanced signal are detected by convolving the same with FOGD operator having 100 ms length and 25 ms standard deviation. After eliminating the spurious peaks, positive and negative peaks of this signal represent locations of VOP and VEP respectively. The region between VOP and VEP represent precise vowel region. Figure 2 depicts the result for a test utterance using FGCI method. Figure 2(a),(b),(c),(d) refers to a continuous speech utterance, Sum of first 3 formant peaks, mean smoothed evidence contour, enhanced evidence using first order Gaussian difference, VOP and VEP marking for the speech signal and prediction respectively.

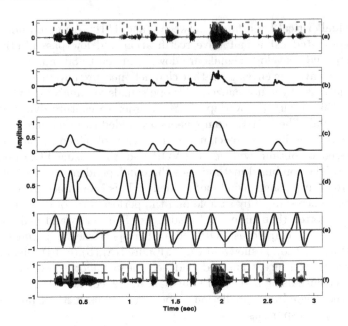

Fig. 2. Vowel region detection method using the FGCI method. (a) speech signal with manually marked vowel regions. (b) sum of 3 formant peaks. (c) mean smoothed evidence contour. (d) enhanced evidence using FOD. (e) VOP and VEP marking for the speech signal. (f) predicted vowel regions.

3   Proposed Vowel Region Detection Method

Zero frequency filtering of a speech signal highlights high information in lower frequency bands [3,12]. VOPs and VEPs are characterized by sudden positive and negative change in energy in low frequency bands respectively. Zero frequency filtering is assumed to be helpful in detecting these acoustic events in a precise manner. In this method, speech signal is passed through a resonator centered at 0 Hz. The resonator is realized using the following transfer function.

$$H(z) = \frac{1}{(1 - z^{-2})^2} = \frac{1}{1 - 2z^{-1} + z^2} \tag{1}$$

The resulting signal is a low pass signal, that exhibits polynomial type growth/delay. This signal is further processed in blocks of 20 ms with a shift of 10 ms. A 256 point DFT is computed and 10 largest peaks are selected in each block. The sum of these spectral peaks is plotted as a function of time. The change at VOP and VEP is perceived at spectral peaks and valleys and it is enhanced by computing its slope using FOD. These enhanced values are convolved with FOGD operator to produce evidence for VOPs and VEP s respectively. The region between a VOP and VEP is considered as a vowel region.

Zero frequency filtered signal is chosen instead of original speech signal as an input to highlight energy in low frequency bands. A sharp rise and fall of

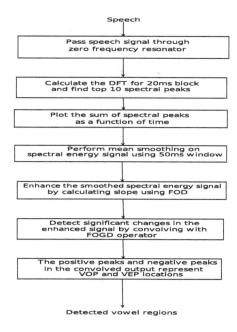

Fig. 3. Flow diagram for the proposed vowel region detection method

energies around glottal closure regions can be visualized as VOPs and VEPs. The
sequence of steps involved in this method are shown in Fig. 3. Experimentation
result for a speech utterance is shown in Fig. 4. Figure 4(a) shows speech signal
along with manually marked vowel regions. Zero frequency filtered signal, sum of
spectral peaks, smoothed contour of spectral peaks, enhanced FOD signal and
FODG operated signal are shown in Fig. 4(b),(c),(d),(e) and (f) respectively.
The vowel region detected by the proposed method is inline with the manually
marked Vowel regions. This is depicted in Fig 4(g). The results produced for the
proposed method is relatively better than COMB method and FGCI methods
in terms of detection rate and false alarms.

4   Performance of the Proposed Method

The proposed approach is evaluated by considering a subset of TIMIT database.
100 test utterances from TIMIT dataset, spoken by 25 speakers (15 male and
10 female) are used for evaluating vowel detection method from a continuous
speech. The detected vowel regions are compared with manually marked vowels
by phonetician. The phonetic transcriptions provided for the datasets are used
as ground truth for vowel region marking. The region between VOP and VEP
in a continuous speech is considered as a vowel region. The Performance of this
method is evaluated based detection and false alarm with percentage overlap of
vowel region with ground truth. Detection refers to a number of vowel regions
identified correctly in-line with ground truth and false alarm correspond to the

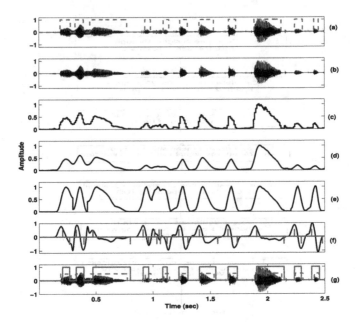

Fig. 4. Vowel region detection method using the proposed method. (a) speech signal with manually marked vowel regions. (b) ZFF signal. (c) sum of 10 spectral peaks in ZFF signal. (d) mean smoothed spectral contour. (e) enhanced spectral energy signal. (f) VOP and VEP marking for the speech signal. (g) predicted vowel regions.

number of non-vowels interpreted as vowels. Table 1 shows the performance of proposed method using TIMIT datasets for clean speech compared with combined method and formant energy based method. The proposed method captures acoustic events and produced relatively improved performance compared with other methods.

Table 2 depicts the performance of the proposed method on TIMIT datasets degraded with additive white Gaussian noise with different signal to noise ratios. It shows that proposed method performs better than other methods in noisy

Table 1. Performance analysis of vowel region detection using combined method (COMB), group delay based formants around GCI (FGCI) and proposed method for a clean speech on TIMIT database. (Det = Detection; FA = False alarm)

Vowel region detection method	Overlap with groundtruth (%)					FA in (%)
	25	50	60	80	90	
	Det	Det	Det	Det	Det	
COMB	84	72	51	24	21	24
FGCI	80	59	25	25	25	10
Proposed	88	75	60	31	31	12

Table 2. Performance analysis of vowel region detection using combined method (COMB), group delay based formants around GCI (FGCI) and proposed method for a noisy speech on TIMIT database. (Det = Detection; FA = False alarm)

SNR 20dB						
Vowel region detection method	Overlap with groundtruth (%)					FA in (%)
	25	50	60	80	90	
	Det	Det	Det	Det	Det	
COMB	84	72	51	24	24	24
FGCI	80	60	51	25	25	10
Proposed	88	74	60	27	24	12
SNR 10dB						
Vowel region detection method	Overlap with groundtruth (%)					FA in (%)
	25	50	60	80	90	
	Det	Det	Det	Det	Det	
COMB	84	71	52	23	24	23
FGCI	80	65	46	23	21	16
Proposed	88	73	60	26	21	12
SNR 5dB						
Vowel region detection method	Overlap with groundtruth (%)					FA in (%)
	25	50	60	80	90	
	Det	Det	Det	Det	Det	
COMB	84	71	52	20	17	24
FGCI	80	63	46	23	21	23
Proposed	88	73	55	25	18	12

environment. It can be highlighted that the robust features can be extracted from zero frequency filtered speech signal.

5 Summary and Conclusions

In this paper, an alternative method for vowel region detection using VOP and VEP was discussed. It was based on the combination of the previous state of art works on VOP and VEP detection. Experimentation was carried out on zero frequency filtered speech signal to highlight the low-frequency content of speech. The performance of the proposed method was compared with recent state of art methods for both clean and noisy speech. The performance of this method was evaluated using TIMIT database and a relative improvement in vowel region detection was observed using proposed method compared to the existing methods. The performance of the proposed method needs to be improved for the test cases with more than 50% overlap with the ground truth.

Future work will aim to improve the performance for higher overlap with ground truths to detect precise vowel regions with less number of false alarms.

References

1. Gangashetty, S.V., Sekhar, C.C., Yegnanarayana, B.: Detection of vowel onset points in continuous speech using autoassociative neural network models. In: Proceedings International Conference Spoken Language Processing, pp. 401–410 (2004)
2. Glass, J.R.: A probabilistic framework for segment-based speech recognition. Comput. Speech Lang. **17**(2), 137–152 (2003)
3. Murty, K.S.R., Yegnanarayana, B.: Epoch extraction from speech signals. IEEE Transa. Audio Speech Lang. Process. **16**(8), 1602–1613 (2008)
4. Prasanna, S.M., Gangashetty, S.V., Yegnanarayana, B.: Significance of vowel onset point for speech analysis. In: Proceedings of International Conference Signal Processing and Communications, pp. 81–88. Citeseer (2001)
5. Prasanna, S.M., Reddy, B.S., Krishnamoorthy, P.: Vowel onset point detection using source, spectral peaks, and modulation spectrum energies. IEEE Transa. Audio Speech Lang. Process. **17**(4), 556–565 (2009)
6. Rao, K.S., Vuppala, A.K.: Non-uniform time scale modification using instants of significant excitation and vowel onset points. Speech Commun. **55**(6), 745–756 (2013)
7. Rao, K.S., Yegnanarayana, B.: Duration modification using glottal closure instants and vowel onset points. Speech Commun. **51**(12), 1263–1269 (2009)
8. Schutte, K., Glass, J.R.: Robust detection of sonorant landmarks. In: INTER-SPEECH, pp. 1005–1008 (2005)
9. Vuppala, A.K., Rao, K.S.: Vowel onset point detection for noisy speech using spectral energy at formant frequencies. Int. J. Speech Technol. **16**(2), 229–235 (2013)
10. Vuppala, A.K., Yadav, J., Chakrabarti, S., Rao, K.S.: Vowel onset point detection for low bit rate coded speech. IEEE Trans. Audio Speech Lang. Process. **20**(6), 1894–1903 (2012)
11. Yadav, J., Rao, K.S.: Detection of vowel offset point from speech signal. IEEE Signal Process. Lett. **20**(4), 299–302 (2013)
12. Yegnanarayana, B., Prasanna, S.M., Guruprasad, S.: Study of robustness of zero frequency resonator method for extraction of fundamental frequency. In: 2011 IEEE International Conference on Acoustics, Speech and Signal Processing (ICASSP), pp. 5392–5395. IEEE (2011)

MiW: An MCC-WMSNs Integration Approach for Performing Multimedia Applications

Joy Lal Sarkar[1]($^{\boxtimes}$), Chhabi Rani Panigrahi[1], Bibudhendu Pati[2], and Rajendra Prasath[3]

[1] Department of Computer Science, Central University of Rajasthan, Ajmer, India
joylalsarkar@gmail.com, panigrahichhabi@gmail.com
[2] Department of Computer Science and Engineering,
C.V. Raman College of Engineering, Bhubaneswar, India
patibibudhendu@gmail.com
[3] Department of Computer and Information Science,
Norwegian University of Science and Technology, Trondheim, Norway
drrprasath@gmail.com

Abstract. The popularity of multimedia applications are growing day-by-day. Emerging Mobile Cloud Computing (MCC) and Wireless Multimedia Sensor Networks (MCC-WMSNs) help to work efficiently with these multimedia applications. The advantage of integrating MCC with WMSNs is that the data gathered by sensor nodes can be accessible from anywhere and at anytime by the users. In this context, there exists several issues such as battery life of the multimedia sensors, processing power, and storage etc. To solve these problems, in this work a MCC-WMSNs integration technique is proposed by using a cloudlets based integration algorithm named as CLIW which mainly works with CLIW-1 and CLIW-2 schemes where both the schemes focus on minimizing energy consumption. The simulation results indicate that CLIW-1 and CLIW-2 schemes are able to prolong the network lifetime of integrated WMSNs.

Keywords: Network lifetime · Multimedia applications · MCC · WMSNs

1 Introduction

The number of smart phone users in India is around 204.1 million and more than 2 billion users in the world and the number is expected to touch 2.7 billion by 2019 [1]. The popularity of using smart phones across the globe is growing due to development of many complex and feature rich applications. In MCC, there are mainly three players: cloud computing, mobile computing, and wireless networks [2]. Cloud computing helps when complex computations are required to perform, where the computation and storage is offloaded by MCC to the cloud.

R. Prasath—A part of this was carried out when the author was in Indian Institute of Information Technology (IIIT) Sricity, India.

R. Prasath and A. Gelbukh (Eds.): MIKE 2016, LNAI 10089, pp. 83–92, 2017.
DOI: 10.1007/978-3-319-58130-9_9

Mobile computing helps to perform the computations in the mobile device itself and hence there is no energy required for offloading purposes [3]. The network is required by the sensor nodes to transfer the data to the mobile for computation purposes. Often the offloading approach is used in MCC, as it may be required when computation is complex and cloud is required to undertake computation and storage activity [4,16]. The results from the cloud are then passed back to the users of mobile devices.

Multimedia aggregates numerous types of contents that are present in varied format and hence dealing with them is quite complex at times. The usage of multimedia started with movies and now spread to other fields such as gaming, medical, education, various computer simulations, home shopping, and training etc. [14,15].

It is of utmost importance that the integration architecture designed should be technically sound and also cater to the applications in various domains (such as banking, telecom, etc.) so that the MCC and WMSNs integration is fully justified. Also, the integration sometimes is a challenge task as both MCC and WMSNs have limited power and therefore it becomes necessary to ensure that energy saving techniques are kept in mind while doing the integration such that the WMSNs devices can function for a longer life and consume energy only when required to reduce the wastage such as in case of Always On (AO) state [5].

In this work, an energy efficient transmission between WMSNs and mobile devices has been proposed named as MiW. The MiW technique provides an efficient connection between sensor nodes and mobile devices. The proposed technique works based on CLIW scheme where CLIW scheme is further divided into CLIW-1 and CLIW-2 schemes. The proposed MiW technique is mainly designed to improve the energy efficiency of multimedia sensor nodes when integrated with MCC. The proposed technique also helps to improve the degree of computation and resources by considering Cloudlets (CLs).

The rest of the paper is organized as follows: Sect. 2 describes the related work. Section 3 describes the proposed MCC-WMSNs integration model along with CLIW scheme. Section 4 presents the simulation obtained along with the analysis of results. Finally, the paper is concluded in Sect. 5.

2   Related Work

In [14], Wang et al. discussed about Cloud Mobile Media (CMM) applications and how offloading done through MCC helps multimedia applications which need computationally intensive and huge network bandwidth for processing. Zhu et al. [6] discussed about the integration of MCC and WSNs in a manner that is energy efficient and at the same time the transmitted data is also valuable. Das et al. [7] considered the case of health monitoring of the patients using sensors that are attached to their body. The resources in the cloud are assigned to processes depending on their expected utilization and priority is given to the process who uses the highest number of resources.

In [8], an integration mechanism has been proposed between different types of sensors which transmit data which is then processed and culminated using

advanced machine learning algorithms and subsequently help out in the irrigation process. Shah et al. [9] discussed about the integration among WSNs, Internet, and cloud in an effective manner that reduces cost and also saves energy. The
authors also suggested an architecture which uses efficient algorithms to send the
message about the patients deteriorating medical condition to the right doctor.

Chen et al. [10] proposed an architecture where the cloud is leveraged for its
vast computational power and storage for coming up with a service which rescues
persons who are in danger and need immediate help. Zhu et al. [11] devised a
MCC and WSNs integration framework which not only increases the life span
of WSNs but also it is intelligent enough to forecast the data flow in a secure
manner.

Zhu et al. [5] proposed a mechanism of MCC and WSNs integration which
assumes that the smart mobile device users would mainly ask for data of vicinity
areas. The authors have put forth two approaches: CLSS-1 where the sensor
gets into active state at a particular location only and in CLSS-2 mainly the
performance factor of the WSNs which is integrated with MCC is taken into
account.

3   Integration Model of MCC-WMSNs

The architecture of MCC-WMSNs integration is shown as in Fig. 1(a) where for
collecting data from WMSNs, the mobile user M_i sends data request to CL. CL
then sends request to the sink node and the sink node then collects data from
the sensor node and sends reply to CL. CL then sends reply to M_i. Figure 1(b)
shows the mobility of the mobile devices where a mobile device moves from one
place to another. When M_i is shifted to another place and it is in the range of
another CL then M_i sends request to the current cloudlet denoted as $CL_{current}$
for collecting the data from previous cloudlet denoted as $CL_{previous}$. If M_i does
not receive the data from the CL when it is in the range of $CL_{previous}$ then M_i
accesses those data from the $CL_{current}$.

3.1   CLIW Scheme

In this section, a cloudlet based integration algorithm CLIW is presented for
MCC and WMSNs. CLIW scheme follows two schemes: CLIW-1 and CLIW-2.
The pseudocodes of CLIW-1 as well as CLIW-2 scheme are described as follows.

CLIW-1 scheme: In CLIW-1 scheme, M_i sends data requests to the CL. The
CL then determines whether the requested data is available in the CL or not. If
the data is not available in the CL then CL sends a flag D to the sink node. The
sink node then broadcasts the flag D to sensor nodes. The sensor nodes which
receive the flag D then they maintain awake state otherwise they go to sleep
state. During awake state, sensor nodes send data to the sink nodes. The sink
node then aggregates the data and sends these to the CL and CL sends the reply
to M_i (steps 1–6 of CLIW-1). Each M_i can move from one place to another. In
such situation, if M_i could not able to collect the requested data from earlier

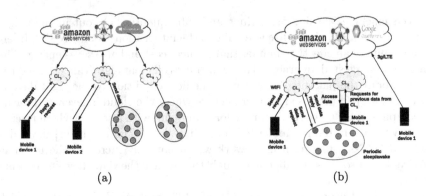

Fig. 1. (a) Architecture of MCC-WMSNs integration, (b) Mobility scenarios of mobile users

CL, then M_i sends request to the $CL_{current}$. The $CL_{current}$ then sends request to the earlier CL for collecting data. The $CL_{current}$ then collects data from the $CL_{previous}$ and sends to M_i. If M_i does not found any CL within its range then it is connected with Central Cloud(CC) (steps 7–9 of CLIW-1).

Algorithm 1. $CLIW$-1 Scheme

1: Mobile device M_i sends data request DR to CL
2: If DR is not available in CL then CL sends a flag D to sink node S_i
3: S_i broadcasts a flag D to sensor nodes
4: If a node n_i receives the flag D, it remains awake and sends reply to S_i otherwise it goes to sleep state
5: S_i aggregates the data and sends reply to the CL
6: CL sends data to M_i
7: If M_i moves to next CL without receiving data then M_i sends request to $CL_{current}$ for accessing data from $CL_{previous}$
8: $CL_{current}$ collects data from $CL_{previous}$ and sends to M_i
9: If there is no CL then M_i connects with CC

CLIW-2 scheme: According to CLIW-2 scheme, M_i sends data request to the CL. The CL then determines whether the requested data is available to the CL or not. If the data is not available to the CL then CL sends a flag D to the sink node.

The sink node replies based on the Residual based Sleep/Awake Scheme (RSAS). According to RSAS, it collects residual energy of 1 to (K-i) nodes ($R^e_{1to(K-i)}$). The nodes can connect to each other either by single hop or multi-hop. A node maintains a flag N if it satisfies two conditions: (a) if $R^e_i > R^e_{1to(K-i)}$ and (b) each node ensures that there are at least K neighbors of it (steps 1–3 of RSAS). CL sends reply to M_i. Each M_i can move from one place to another place. In such situation, if M_i could not be able to collect the requested data

Algorithm 2. CLIW-2 Scheme

1: Mobile device M sends data request DR to CL
2: If DR is not available to CL then CL sends a flag D to sink node S_i
3: S_i sends reply to CL based on RSAS
4: CL sends data to M_i
5: If M_i moves to next CL without receiving data then M_i sends request to $CL_{current}$
   for accessing data from $CL_{previous}$
6: $CL_{current}$ collects data from $CL_{previous}$ and sends to M_i
7: If there is no CL then M_i connects with CC

Algorithm 3. RSAS

1: Collects current R_i^e
2: Node A connects with node B either by direct connection (single hop) or indirect
   connection (multi-hop)
3: Each node A awakes if it receives a flag N and then goes to sleep if
   (a) $R_i^e > R_{1to(K-i)}^e$
   (b) Each node ensures that there are atleast K neighbors of it

from $CL_{previous}$, then M_i sends request to the $CL_{current}$. The CL then sends request to the $CL_{previous}$ for collecting data. The $CL_{current}$ then collects data from the $CL_{previous}$ and sends to M_i. If M_i does not find any CL within the range of M_i then it connects with CC (steps 4-7 of CLIW-2).

3.2 Analysis of CLIW Scheme

We assume that in a 2-D area χ, N number of nodes are deployed in WMSNs. The node density denoted by ψ can be computed by using Eq. (1).

$$\psi = \frac{N}{\chi} \tag{1}$$

Now, from each n_i the distance of the nearest neighbor [12,13] denoted by $d(n_i, N_n)$ is computed by using Eq. (2).

$$d(n_i, N_n) = \frac{1}{2\sqrt{\psi}} \tag{2}$$

Let T be the time interval for node n_i when the expected number of events occur in WMSNs denoted by T^e and is computed by using Eq. (3).

$$T^e = \Psi.u \tag{3}$$

where, u denotes the probability that an event occurs.

Now, assuming that there are p_i number of packets that are transmitted from n_i after detection of an event. So, for each n_i, the total number of packets (P) that are transmitted can be computed by using Eq. (4).

$$P = \psi.\pi.T^2.p_i.\Psi.u \tag{4}$$

Now, for each sensor, the expected number of neighbors denoted by $exp(n)$ can be computed by using Eq.(5).

$$exp(n) = \psi.\pi.T^2 \tag{5}$$

Each sensor node maintains both sleep as well as awake state periodically under CLIW-1 scheme. If a sensor node does not receive any flag from CL then that sensor node gets back to the sleep state. So, the energy consumed by each n_i denoted by E_c^{CLIW1} and is computed by using Eq. (6) [5].

$$E_c^{CLIW1} = E^t.(\psi.\pi.T^2) + E^r + \frac{P(E^t + E^r)}{2} - \omega(CC) \tag{6}$$

In Eq. (6), $\omega(CC)$ denotes the extra communication cost which is considered as energy consumption due to high latency. The negative sign indicates that most of the time CLIW-1 ignore CC when CL is available. Here, E^t and E^r denote the consumption of energy for transmitting and receiving packets respectively. So, the network lifetime in CLIW-1 denoted as Ne(life1) and can be computed by using Eq. (7).

$$Ne(life1) = \frac{Io}{E^t.(\psi.\pi.T^2) + E^r + \frac{P(E^t+E^r)}{2} - \omega(CC))}.T \tag{7}$$

where, I_o denotes the initial energy of each sensor node.

Now, according to CLIW-2 scheme, the nodes are awake when they receive flag N from neighboring nodes. In such scenario, the nodes consume their energy. CLIW-2 scheme follows $RSAS$. Under $RSAS$, sensor nodes can be in awake state after satisfying both the conditions (a) and (b). So, each n_i consumes its energy under CLIW-2 scheme denoted by E_c^{CLIW2} and can be computed by using Eq. (8).

$$E_c^{CLIW2} = E^t.(\psi.\pi.T^2) + E^r + \frac{P(E^t + E^r)}{2} + \frac{(E^t + E^r)P^*}{2} - \omega(CC) \tag{8}$$

where, P^* denotes the total number of transmitted packets. The network lifetime in CLIW-2 denoted as Ne(life2) and can be computed by using Eq. (9).

$$Ne(life2) = \frac{\frac{I_o}{E^t.(\psi.\pi.T^2)+E^r+\frac{P(E^t+E^r)}{2}+\frac{(E^t+E^r)P^*}{2}} - \omega(CC)}{2}.T \tag{9}$$

The CL has limited resources and for that CL executes other jobs after execution of its own job and for that we use External Service Ratio (ESR) [14].

Theorem 3.1. *The network lifetime under CLIW-1 scheme is equal to the product of network lifetime of CLIW-2 and $(2 + \frac{P^*}{p^*})$ if energy consumption during transmission is too high.*

Proof. The ratio of the network lifetime between two schemes is given in the following:

$$\frac{Ne(life1)}{Ne(life2)} = \frac{Io}{Et.(\psi.\pi.T^2) + E^r + \frac{P(E^t+E^r)}{2} - \omega(CC)}.T$$

$$\div \frac{\frac{I_o}{Et.(\psi.\pi.T^2)+E^r+\frac{P(E^t+E^r)}{2}+\frac{(E^t+E^r)P^*}{2}-\omega(CC)}}{2}.T$$

$$= \frac{Io}{Et.(\psi.\pi.T^2) + E^r + \frac{P(E^t+E^r)}{2} - \omega(CC))}$$

$$\times \frac{2.\left(Et.(\psi.\pi.T^2) + E^r + \frac{P(E^t+E^r)}{2} + \frac{(E^t+E^r)P^*}{2}\right) - \omega(CC)}{I_o}$$

$$= \frac{2.\left(Et.(\psi.\pi.T^2) + E^r + \frac{P(E^t+E^r)}{2} + \frac{(E^t+E^r)P^*}{2}\right) - \omega(CC)}{Et.(\psi.\pi.T^2) + E^r + \frac{P(E^t+E^r)}{2}) - \omega(CC)}$$

$$= \frac{2.Et.(\psi.\pi.T^2) + 2E^r + P(E^t + E^r) + (E^t + E^r)P^* - 2\omega(CC)}{Et.(\psi.\pi.T^2) + E^r + \frac{P(E^t+E^r)}{2}) - \omega(CC)}$$

$$= \frac{E^t\left(2\psi.\pi.T^2 + 2.\frac{E^r}{E^t} + P(1 + \frac{E^r}{E^t}) + P^*(1 + \frac{E^r}{E^t}) - \frac{2\omega(CC)}{E^t}\right)}{Et.(\psi.\pi.T^2) + E^r + \frac{P(E^t+E^r)}{2}) - \omega(CC)}$$

$$= 2.\frac{E^t\left(2\psi.\pi.T^2 + 2.\frac{E^r}{E^t} + P(1 + \frac{E^r}{E^t}) + P^*(1 + \frac{E^r}{E^t}) - \frac{2\omega(CC)}{E^t}\right)}{E^t\left(2.\psi.\pi.T^2 + 2.\frac{E^r}{E^t} + P(1 + \frac{E^r}{E^t}) - \frac{2\omega(CC)}{E^t}\right)}$$

$$= 2 + \frac{2.P^*(1 + \frac{E^r}{E^t})}{\left(2.\psi.\pi.T^2 + 2.\frac{E^r}{E^t} + P(1 + \frac{E^r}{E^t}) - \frac{2\omega(CC)}{E^t}\right)}$$

$$When \ E^t >> E^r \ and \ \omega(CC)$$

$$= 2 + \frac{2.P^*}{(2.\psi.\pi.T^2 + P)}$$

$$= 2 + \frac{P^*}{p^*}; \ Where, \ p^* = \left(\psi.\pi.T^2 + P/2\right)$$

$$So, Ne(life1) = Ne(life2)(2 + \frac{P^*}{p^*})$$

Therefore, it is clear that the network lifetime under CLIW-1 scheme is equal to the product of network lifetime of CLIW-2 and $(2 + \frac{P^*}{p^*})$.

4   Simulations

MiW was evaluated based on android operating system. Android x86 was installed on Intel I3 laptop. Samsung I997 was used for deploying the applications. We run our approach in a stand alone environment where unwanted applications are completely closed and background jobs are also shutdown. The static and dynamic power of CPU was set from 0.3 to 1 randomly and also the clock frequency was set ranging from 1.2 GHz to 1.6 GHz randomly. The energy consumption depends on the wireless connectivity i.e. the good connectivity saves more energy then the bad connectivity. The mobile phones use only one connection at a time and the first priority was given for WiFi. We used MAT-LAB for developing WMSNs model where network area considered was 400 × 400 m^2 and the nodes were deployed randomly ranging from 100 to 1000. The average rate of event occurs was assumed to be 35 times/minute. The initial energy of each node was set to 3.1 J. The consumption of energy during transmitting and receiving packets were set to 0.0144 mJ and 0.00576 mJ respectively. The coverage range of each cloudlet considered was 15 m and the density was set to 0.0003 per m^2. The mobile users were located at the cloudlets coverage area and the time for each mobile user located at one cloudlet range and was considered as 5 m. During this time, if a mobile user does not receive data reply from cloudlets, the mobile user sends data request from the current cloudlet to the earlier cloudlet.

4.1 Performance Analysis

Figures 2 and 3 show the analysis of lifetime of the network of CLIW scheme and the comparison with two baselines that are AO and $CLSS$ scheme for different mobile users. From Figs. 2 and 3, it is observed that the MCC-WMSNs integration enhances the network lifetime greatly as compared to two baselines AO and $CLSS$ schemes. Figures 2 and 3 indicate that the network lifetime is more in case of CLIW scheme in comparison with two baselines AO and $CLSS$ schemes. Figures 2 and 3 also indicate that the network lifetime in CLIW-1 scheme is longer as compared to CLIW-2 scheme. The proposed CLIW-2 performs better as compared to the AO scheme and this is because AO scheme considers that the sensor nodes always in the awake state where as sensor nodes maintain sleep scheduling technique in case of CLIW scheme. On the other hand, both CLIW

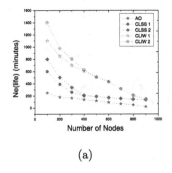

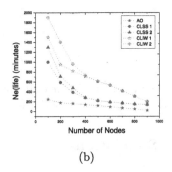

(a)   (b)

Fig. 2. Lifetime of the network under CLIW, CLSS, and AO scheme for mobile user1 (a) and user2 (b).

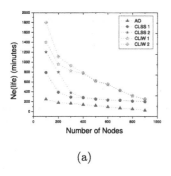
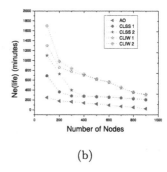

(a)   (b)

Fig. 3. Lifetime of the network under CLIW, CLSS, and AO scheme for mobile user3 (a) and user4 (b).

and *CLSS* scheme consider the sleep scheduling technique but in later case the sensor nodes always depend on the location of the mobile devices. On the basis of a user request, the nodes awake and successfully respond by considering the public cloud as a middle layer. Consequently, the mobile users are able to receive sensory data through the public cloud. As compared to the sensor-cloud connectivity, the sensor nodes consume less energy when connected with the cloudlets. Proposed CLIW scheme is fully utilized when the mobile device and the sensor nodes are connected via cloudlets.

5 Conclusion

In this work, an MCC-WMSNs integration approach has been proposed which is based on the CLIW-1 and CLIW-2 scheme. The simulation results indicate that the network life time is longer in case of CLIW-1 scheme as compared to CLIW-2 scheme. The simulation results also indicate that both CLIW-1 and CLIW-2 schemes are able to prolong the network lifetime in an integrated WMSNs.

References

1. http://www.statista.com/statistics/467163/forecast-of-smartphone-users-in-india/. Accessed 1 Sep 2016
2. https://en.wikipedia.org/wiki/Mobile_cloud_computing. Accessed 1 Sep 2016
3. https://en.wikipedia.org/wiki/Wireless_sensor_network. Accessed 1 Sep 2016
4. Kumar, K., Liu, J., Lu, Y.-H., Bhargava, B.: A survey of computation offloading for mobile systems. Mobile Netw. Appl. **18**(1), 129–140 (2012)
5. Zhu, C., Leung, V.C.M., Yang, L.T., Shu, L.: Collaborative location-based sleep scheduling for wireless sensor networks integrated with mobile cloud computing. IEEE Trans. Comput. **20**(20), 1–14 (2014)
6. Zhu, C., Sheng, Z., Leung, V.C.M., Shu, L., Yang, L.T.: Toward offering more useful data reliably to mobile cloud from wireless sensor network. IEEE Trans. Emerg. Top. Comput. **3**(1), 84–94 (2015)

7. Das, S., Misra, S., Khatua, M., Rodrigues, J.: Mapping of sensor nodes with servers in a mobile health-cloud environment. In: Proceedings of IEEE 15th International Conference on e-Health Networking, Applications & Services (Healthcom) (2013)

8. Li, C., Dutta, R., Kloppers, C., D'Este, C., Morshed, A., Almeida, A., Das, A., Aryal, J.: Mobile application based sustainable irrigation water usage decision support system: an intelligent sensor cloud approach. In: SENSORS (2013)

9. Shah, S.H., Iqbal, A., Shah, S.S.A.: Remote health monitoring through an integration of wireless sensor networks, mobile phones & cloud computing technologies. In: Proceedings of IEEE 2013 Global Humanitarian Technology Conference (2013)

10. Chen, Y.-J., Lin, C.-Y., Wang, L.-C.: Sensors-assisted rescue service architecture in mobile cloud computing. In: IEEE Wireless Communications and Networking Conference (WCNC) (2013)

11. Zhu, C., Wang, H., Liu, X., Shu, L., Yang, L.T., Leung, V.C.M.: A novel sensory data processing framework to integrate sensor networks with mobile cloud. IEEE Syst. J. $10(3)$, 1125–1136 (2014)

12. Wang, L., Yuan, Z., Shu, L., Shi, L., Qin, Z.: An energy-efficient ckn algorithm for duty-cycled wireless sensor networks. Int. J. Distrib. Sensor Netw. $8(5)$, 1–15 (2012)

13. Zhu, C., Yang, L.T., Shu, L., Hara, T., Nishio, S.: Implementing top-k query in duty-cycled wireless sensor networks. In: Proceedings 7th International Conference Wireless Communications Mobile Computing Conference, pp. 553–558 (2011)

14. Wang, S., Dey, S.: Adaptive mobile cloud computing to enable rich mobile multimedia applications. IEEE Trans. Multimedia $15(4)$, 870–883 (2013)

15. Wang, S., Dey, S.: Modeling and characterizing user experience in a cloud server based mobile gaming approach. In: Proceeding of IEEE GLOBECOM (2009)

16. Panigrahi, C.R., Pati, B., Tiwary, M., Sarkar, J.L.: EEOA: improving energy efficiency of mobile cloudlets using efficient offloading approach. In: Proceedings of IEEE International Conference on Advanced Networks and Telecommunications Systems, pp. 1–6 (2015)

An Allocation Technique of MMH/FH
for an Aircraft

Antony Gratas V.$^{(\boxtimes)}$ and Prakash R.

Aeronautical Development Agency, Bangalore 560017, India
vaagratus@yahoo.co.in

Abstract. Maintenance of any item during its life cycle plays vital role in determining the effectiveness of the item to customer and the total life cycle cost associated with it, in the long run, till a decision on discard is accorded. It is imperative to minimize the MDT [Mean Down Time] of any item/system so as to achieve the on-demand availability, as high as possible. Designing for maintainability in coherence with functionality and reliability requirements would eventually determine the operational effectiveness of the item. Especially, in the case of fighter aircraft, where availability is required to be ensured to the maximum possible, which is one of the critical factors in determining the battle strike capability of the armed forces. In other words, Maintenance Man Hour per Flight Hour [MMH/FH] is required to be as low as possible, so as to ensure the on-demand availability of the fighting platform. In this aspect, every stages of the design should cater for the target MMH/FH. The MMH/FH is required to be allocated from the user requirements to the systems, through sub-systems to the LRUs. This paper is aimed at, presenting one of the methodologies to allocate the MMH/FH goal to the systems/subsystems/LRUs, considering all the applicable factors of influence, during the design stage. The proposed Allocation Model has been validated, using a typical fighter aircraft, ensuring correctness and completeness.

Keywords: Allocation · Availability · Diagnostics · Failure rate · Fighter aircraft · Maintainability · Mean down time · MMH/FH · MSF · MTBF · MTTF · Prognostics · Reliability

1 Introduction

Ensuring on-demand availability of the fighter aircraft requires very minimal maintenance efforts, which is expressed in terms of MMH/FH. It is required to ensure the availability of the aircraft, with no or minimal repair for the next mission, upon completion of a mission in a day or in the subsequent days. MMH/FH is expressed in terms of maintenance man hours required to ready the aircraft. The preparation of the aircraft for the stated mission requires different levels of maintenance. A typical fighter aircraft is considered for this approach and a methodology has been derived on the approach by which the MMH/FH could be designed into the system. Considering the user requirements as the

© Springer International Publishing AG 2017
R. Prasath and A. Gelbukh (Eds.): MIKE 2016, LNAI 10089, pp. 93–104, 2017.
DOI: 10.1007/978-3-319-58130-9_10

goal for the MMH/FH, this paper aims at identifying all the critical parameters which would have significant influence on the system availability and hence the MMH/FH. A mathematical formulation has been derived for the optimal allocation of MMH/FH for the constituent systems/sub-systems/LRUs of the aircraft. The methodology proposed here in this paper is first off kind among the literatures for the purposes allocation of MMH/FH, excluding the cost constraints, though implicitly resulting into reduction of LCC. Allocation is usually carried out in multiple levels, in a tree structure as shown in the following indenture levels. Refer Fig. 1.

- Aircraft Level MMH/FH Goal
- System Level MMH/FH Goal
- Sub-system Level MMH/FH Goal
- LRU Level MMH/FH Goal

In this paper, we have formulated and enumerated a mathematical model to derive the values of $X_{i,j,k}$ from a given X. Where,

$$a : \text{No. of systems, given by } i$$
$$b : \text{No. of sub-systems, given by } j$$
$$c \cdots e : \text{No. of LRUs in subsytem } 1 \cdots \text{ b, given by } k$$

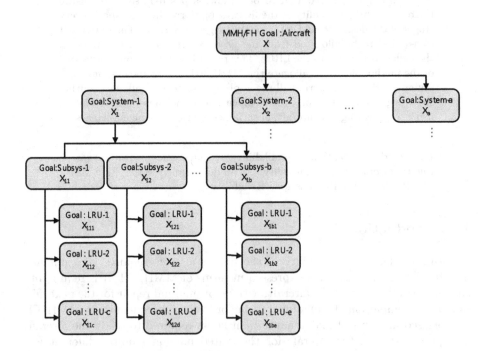

Fig. 1. Levels of MMH/FH allocation

2   Maintenance and Types

Maintenance is often dictated by the Mean Time to Repair, primarily. In accordance with the definition of maintenance, it is the total time required to bring back the system into operational state from non-operational state, when the maintenance procedures specified are adhered to. They are represented in time domain and presented in Fig. 2. In theoretical terms the amount of maintenance efforts required are predominantly determined by the failure rate of the system. In other words, an item with the higher failure rate is expected to fail more frequently during the life cycle and hence it is very much essential to design-in features, which would ensure the MTTR as low as practically possible, so as to maintain/optimize the availability. Refer [1]. Expressing mathematically,

$$MTTR_{(i)} = \frac{MTTR_{(G)}.MTBF_{(i)}}{n} \sum_{i=1}^{n} \frac{1}{MTBF_{(i)}} \qquad (1)$$

where,

$MTTR_{(i)}$: MTTR of item i
$MTTR_{(G)}$: Goal of MTTR for the system
$MTBF_{(i)}$: MTBF of item i
n : Total number of items in the system

There are two types of maintenance philosophy adopted, in general, in the aircraft industry. Those are categorized based on the types of failure mechanisms and the occurrence rates. These could be deterministic or probabilistic, based on the nature of failures and repair/replacements.

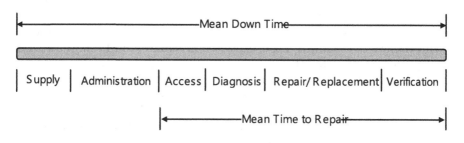

Fig. 2. Maintenance times

2.1   Scheduled Maintenance

All items which require periodic maintenance in terms on CLAIR [Clean, Lubricate, Analyse, Inspect & Repair] are categorized under preventive maintenance or scheduled maintenance category. Moving/rotating components are required to be maintained, irrespective of the usage and occurrence of failures. Maintenance efforts for these items are required to be minimized, if not, fully eliminated, with the aid of advanced health and usage monitoring. Efforts required and the frequency of maintenance of these kind increases as the age of the item under consideration increases. Refer-[3].

2.2 Unscheduled Maintenance

Categories of items/components which exhibits failures without any symptoms and with randomness in every occurrence and type of failure, could largely be categorized under unscheduled maintenance classification. Electronics based hardware and software components, where the fault propagation is triggered by many random factors are treated under this category. However, continuous monitoring with in-built software/firmware would eventually render alarms upon occurrence of failures. With suitable sensing mechanism of critical parameters, the health of the electronic based system could be monitored and warning could be generated well ahead of the occurrence of the failures. Hence, the maintenance actions required are very minimal for these cases, as repair time is very minimal. Frequency of occurrence of these kind of failures is also expected to very less and hence the maintenance efforts. Using PHM techniques, the efforts and hence the total cost required to maintain the item, could be significantly reduced. Refer-[6].

3 Maintenance Man Hour per Flight Hour [MMH/FH]

As referred in Sect. 2, it is very well evident that, both scheduled and unscheduled maintenance are required to be considered while drawing the maintenance requirements for the system. Apart from MTTR, the mean man power required to carry out any maintenance task, within the stipulated time, plays a vital role in deriving the effectiveness of maintenance program. As the MTTR increases, the MMH required to carry out any given task will also increase. MMH/FH is a dimensionless number, which determines the availability in terms of maintenance man hours required to prepare the aircraft for flying. In other words, it is the average man power required to resolve/repair the task assigned. Although the effectiveness of the MMH/FH is expressed in terms of MTTR in the aircraft industry, there are many other factors which would significantly increase the MMH/FH. In order to improve the MMH/FH, it is required to ensure that, the MTTR is as low as possible, while controlling other significant factors of MDT. It is mathematically evident from the Eq. 1 that, the MTTR is determined as a function of MTBF, wherein the MTTR of an item, which has the higher failure rate should have the lowest possible MTTR, in order to achieve the required availability.

4 Factors Influencing MMH/FH

In order to appreciate the MMH/FH, it is very much essential to figure out the underlying factors, which would significantly affect the MMH requirement of an aircraft. There are numerous activities which are required to be carried out prior to arrive at the inherent MTTR of any item. The following factors play vital role in determining the MMH of the aircraft.

- Accessibility
- Diagnostics [Testability]
- Prognostics
- Complexity
- Fault Tolerance
- Maintenance Frequency

Significance of the above listed factors are explained in detail in the following subsections.

4.1   Accessibility

In a complex platform like a fighter aircraft, where the available space for installing any item is very less, accessibility is one of the critical parameters. The main requirement of having all the functionalities, equipment and interfaces on-board, it is impossible to ensure that, all the items are accessible. However, based on the expected failure rate of any item, and its occurrence of failures in real time, the item could be installed in such a way as to optimize the availability. More specifically, items with very high failure rate are expected to fail more frequently, than the items with lesser failure rate. As a result, items with high probability of failure is required to be more accessible and with ease. This is a significant portion of time in the MMH/FH, as the items which are easily accessible for installation, removal and diagnostics, would eventually have lesser MMH/FH.

4.2   Diagnostics [Testability]

For the purposes of corrective maintenance, testability feature adds more value in declaring the identified fault, isolating, confining and error recovery. It aids in reducing the maintenance time and efforts. An item with good BIT (Built-In-Test) facility, is desired for, for better maintenance management. Testability covers all the components, if not, critical components in an item, whose failure would significantly reduce the effectiveness of the system, which may result in down time. Each item shall have sufficient test points for the measurement of internal design to achieve the capability of detecting failures using full resources. Automatic monitoring and diagnostic capabilities shall be provided to show the system status (operable, inoperable, and degraded) and to detect failures. This factor is influential in positively reducing the MTTR, to a significant level.

4.3   Prognostics

Continuous monitoring of the critical parameters of life of an item, thereby forecasting the useful remaining life and the expected time to failures are of the major outcomes of *'Prognostic Maintenance'*. Maintenance actions required to be carried out, will be known *a-priori* and accordingly resources could be optimized ensuring lesser time for repair and/or replacement. When prognostics feature is built-in, within the item design and incorporated, the maintenance efforts required would drastically be lowered, which is explained in [5,6]. The maintenance actions required could be planned in advance, using the prognostics feature, and thereby the occurrence of unexpected maintenance actions could be reduced significantly.

4.4   Complexity

Increase in complexity directly affects and effects in the reliability and availability of the system. Irrespective of the state-of-the-art technology adaptation, the maintenance involves training and specialization in the cutting-edge technology and expertise and proportionately increases the MMH/FH of the system. Higher the complexity of an item, higher would be skill level required to maintain the item.

4.5   Fault Tolerance

Whilst a graceful degradation is permitted in, as for as the functional requirements are considered, the failures warranting maintenance actions may be '*delayed*', based on the acceptable operable limits. Hence, the allocation of MMH/FH could be lesser to those systems with '*fault tolerant*' capability, than those systems without it, under the conditions that, the occurred failures does not affect/hamper continuous and safe operation, in presence of faults. Fault Tolerance is the design capability aims primarily to handle acceptable faults in terms of functionality ensuring availability, and secondarily to reduce the maintenance actions required.

4.6   Maintenance Frequency

For the case of scheduled maintenance, the frequency within which the operations are carried out determines the major portion of MMH/FH, as these maintenance actions are dictated by life characteristics of the system/LRUs. Those items which require periodic inspection, for example, tyre pressure, fuel level, hydraulic fluid pressure fall under this category. Maintenance frequency β, could be reduced, if prognostics feature in-built on these life limited items.

The factor β_i, which is applicable only for scheduled maintenance is determined by the fraction of time and/or frequency by which the unscheduled maintenance activity is expected to be executed. Expressed mathematically,
For calendar based maintenance,

$$\beta_i = \frac{t_i}{T} \tag{2}$$

For usage based maintenance,

$$\beta_i = \frac{m_i}{M} \tag{3}$$

where,

T : Total duration of operation, prior to scheduled maintenance
t_i : Total duration of maintenance of an item i, during T
M : Total calendar time of operation, prior to scheduled maintenance
m_i : Expected time to repair in calendar time of item i, during M

5   Allocation Methodology for MMH/FH

Design for MMH/FH is also one among the design constraints during the design trade-off studies. In order to conceive the MMH/FH, during the design cycle, the goal of MMH/FH needs to be specified by the customer. If not, the same could be captured using '*Quality Function Deployment [QFD]*'. Once the goal for MMH/FH is certain and known, the same needs to be further reapportioned to the next lower indenture levels. During this process of allocation of aircraft level requirements down to the item level MMH/FH, through system level and sub-system level, there are various factors, which would have influence on the MMH/FH, which are to be considered and appropriately modelled. In the present proposal of allocation of MMH/FH of a typical fighter, many indenture levels have been considered, as given in Sect. 1. There are basically two major types of maintenance in aircraft industry. They are,

– Unscheduled Maintenance
– Scheduled Maintenance

The relationship of MTBF is very well known with the fact that, the failure rate would be decreased to a significant level, while reducing the duty cycle, as similar to de-rating of the electronic components. Accordingly, MTBF is expected to increase, but not beyond the design limit. An equipment or item which is designed to operate with 100 percentage duty cycle, is expected to function longer than the design limit, if operated at significantly lower duty cycle. Proportionately the MMH/FH required to maintain this item, over the life cycle would also be lesser. Accordingly, effective MTBF is expressed as,

$$EMTBF_i = \frac{MTBF_i}{DC_i} \qquad (4)$$

where,

$EMTBF_i$: Effective MTBF for item i
DC_i : Duty cycle for item i, ranging from 0 to 1
$MTBF_{(i)}$: MTBF of item i

The methodology adopted in defining the MMH/FH allocation technique for both the above types of maintenance are explained in the forthcoming sections.

5.1 MMH/FH for Unscheduled Maintenance

Unscheduled maintenance are typically applicable for those items/systems whose failure distributions are randomized. Typically to quote, the exponential probability density function is most suited for defining the failure phenomena of electronics based systems/LRUs, whereas, the Weibull *pdf* is more suited for the mechanical system/LRUs. The MMH/FH for the case of unscheduled maintenance is modelled as a function of *Effective MTBF* and *MSF*, which is given below,

$$f(MMH/FH_{i-USCH}) \propto \frac{\dfrac{\theta_{Ui}}{EMTBF_i}}{\displaystyle\sum_{i=1}^{n} \dfrac{\theta_{Ui}}{EMTBF_i}} \tag{5}$$

$$f(MMH/FH_{i-USCH}) = \frac{\dfrac{\theta_{Ui}}{EMTBF_i}^{\alpha_i}}{\displaystyle\sum_{i=1}^{n} \dfrac{\theta_{Ui}}{EMTBF_i}} \tag{6}$$

And

$$\theta_{Ui} = \frac{C_i}{A_i.D_i.FT_i} \tag{7}$$

where,

$$
\begin{aligned}
USCH &: \text{Unscheduled Maintenance} \\
MMH/FH_{i-USCH} &: \text{MMH/FH for item } i, \text{ considering USCH} \\
\alpha_i &: \text{Constant of proportionality for USCH for item } i \\
n &: \text{Total number of items in the system} \\
\theta_{Ui} &: \text{Maintenance Significant Factor [MSF] for item } i, \\
&\quad \text{ for USCH} \\
C_{(i)} &: \text{Complexity factor of item } i \\
A_{(i)} &: \text{Accessibility factor for of item } i \\
D_{(i)} &: \text{Diagnostics factor for item } i \\
FT_{(i)} &: \text{Fault Tolerant factor of item } i
\end{aligned}
$$

The above factors are chosen based on the system architecture, installation and design features, on the common whole number scale given that $1 < A_i, C_i, D_i, FT_i < 5$, where 1 being the least and 5 being the most on the chosen factor.

The factor α_i is being chosen in such a way as to represent the testability of the failure occurred. It ranges from 0.1 to 0.5. When the testability is 100%, for any reported failure, a factor of 0.5 is assumed and if 0%, 0.1 is assumed.

5.2 MMH/FH for Scheduled Maintenance

As referred in Subsect. 2.1, the scheduled maintenance is determined while the architecture of the system is known. Data from legacy platform/similar item or from manufacturer are taken into consideration, while establishing the deterministic value. Periodicity, though well-defined and determined could further be optimized based on prognostic feature, if built-in. Refer [2]. The man hours that are expected to be spent for scheduled maintenance depends on the frequency of the maintenance actions and other applicable maintenance factors.

$$f(MMH/FH_{i-SCH}) \propto \frac{\theta_{Si}}{\sum_{i=1}^{n} \theta_{Si}} \tag{8}$$

$$f(MMH/FH_{i-SCH}) = \frac{\beta_i \cdot \theta_{Si}}{\sum_{i=1}^{n} \theta_{Si}} \tag{9}$$

$$\theta_{Si} = \frac{C_i}{A_i.D_i.FT_i} \tag{10}$$

where,

MMH/FH_{i-SCH} : MMH/FH for item i, considering scheduled maintenance
β_i : Constant of proportionality for SCH maintenance for item i
i.e., Maintenance Frequency
n : Total number of items in the system
θ_{Si} : MSF for item i, for scheduled maintenance
FT_i : 1, for scheduled maintenance [No fault tolerance]

5.3 MMH/FH for Total Maintenance

From the previous Subsects. 5.1 and 5.2, the functional relationship of unscheduled maintenance and scheduled maintenance with various applicable parameters and factors have been established. In order to allocate the target MMH/FH to the next level, the total maintenance efforts are to be accounted and normalized. As the fraction of MMH/FH which would be allocated to the systems with scheduled maintenance and unscheduled maintenance are different, separate scheme of interrelationships have been drawn in Eqs. 6 and 9.

While combining, we get,

$$
\left[\frac{MMH}{FH}\right]_i = \left[\frac{MMH}{FH}\right]_t \left(\frac{\dfrac{\dfrac{\theta_{Ui}}{EMTBF_i} \cdot \alpha_i}{\displaystyle\sum_{i=1}^{n} \dfrac{\theta_{Ui}}{EMTBF_i}} + \dfrac{\beta_i \cdot \theta_{Si}}{\displaystyle\sum_{i=1}^{n} \theta_{Si}}}{\displaystyle\sum_{i=1}^{n} \dfrac{\dfrac{\theta_{Ui}}{EMTBF_i} \cdot \alpha_i}{\displaystyle\sum_{i=1}^{n} \dfrac{\theta_{Ui}}{EMTBF_i}} + \dfrac{\beta_i \cdot \theta_{Si}}{\displaystyle\sum_{i=1}^{n} \theta_{Si}}} \right)
\tag{11}
$$

where,

$$
\left[\frac{MMH}{FH}\right]_i : \text{MMH/FH for item } i
$$

$$
\left[\frac{MMH}{FH}\right]_t : \text{Target MMH/FH}
$$

6   Validation of the Methodology

In order to validate the devised methodology [named as *VAAG MMH/FH Allocation Technique*], a typical fighter aircraft has been considered with the assumed MMH/FH goal of 10. Allocation of this goal to next identified system levels are taken up. The fighter aircraft is assumed to be constituted by many systems, which are listed in Table 1. The table also summarizes the derived values, and allocated MMH/FH to the systems under consideration. It is assumed in general, that skilled man power would be employed for the maintenance activities, and hence the same has not been accounted for. It is also assumed that the required test equipment, spares and tools are available within the inventory.

MTBF is the critical parameter in the process of allocation of MMH/FH to the next indenture levels. It is assumed that, during the design stages, the target MTBF of the aircraft has been apportioned to the all the systems of the aircraft using available Reliability/MTBF allocation techniques in the literature.

For the validation of the proposed model, in addition to MTBF, other factors as mentioned in Subsect. 5.1 are to be chosen based on the system architecture, installation and design features, on the common whole number, where 1 being the least and 5 being the most on the chosen factor. From Table 2, it is observed that, the allocated MMH/FH is more for those systems, which required to have preventive maintenance. The ideal methodology of reducing the MMH/FH for the typical fighter aircraft would be to instrument the systems reliably, so as to eliminate, if not, to reduce frequency of preventive maintenance checks. Implementation of prognostics would exponentially reduce the down time. The MMH/FH are to be allocated optimally among the systems, considering the applicability of scheduled maintenance and unscheduled maintenance. Refer [4].

While realizing the system, the designers are to be provided with the following design objectives/requirements, in addition to the functional requirements.

Table 1. Maintenance significant factors

Systems	Complexity	Accessibility	Diagnostics	Fault tolerance
Structures	2	5	1	1
Flight control	4	3	4	4
Avionics	4	3	4	5
Power plant	3	2	3	1
Environmental control	3	3	2	1
Electrical power	3	2	4	2
Hydraulics power	3	3	2	2
Fuel system	3	2	2	1
Safety system	2	3	1	1
Landing system	4	4	1	1

- MTBF goal
- MSF goals
- MMH/FH goal

Both MSF and MMH/FH Target would be given to the designer to realize the same, during early stages of design. When the assumed MSF is found to be impossible to realize/conceive within the system design, the MMH/FH goals for all the other systems would get changed, so as to meet the target MMH/FH of aircraft. If all the systems are in the infancy stage of design and development, an iteration(s) could be initiated to study trade-off and feasibility study of desired MSF. The chosen design option shall comply with the stated requirements, however, at a later stage, if the targets assigned are not met, then target MMH/FH of aircraft would not be met.

Table 2. Validation of the allocation model

Systems	MTBF	θ_{Ui}	α_i	θ_{Si}	β_i	$\left[\frac{MMH}{FH}\right]_i$
Structures	11000	0.4	0.2	0.4	0.2	0.22424
Flight control	2000	0.0833	0.5	0.333	0.2	0.25805
Avionics	1100	0.0667	0.5	0.333	0.1	0.22485
Power plant	1200	0.5	0.2	0.5	0.4	0.87455
Environmental control	1400	0.5	0.3	0.5	0.7	1.34403
Electrical power	2000	0.1875	0.5	0.375	0.1	0.31456
Hydraulics power	2700	0.25	0.3	0.5	0.7	0.96160
Fuel sytem	1400	0.75	0.3	0.75	0.8	2.19343
Safety system	1800	0.667	0.1	0.667	0.4	0.80913
Landing system	1600	1	0.3	1	0.8	2.79550

7   Conclusion

Mathematical formulation in assigning the MMH/FH target to the lower indenture levels, considering the typical fighter aircraft has been carried out. Detailed investigations into the system architecture, enabled in assigning the Maintenance Significant factors. Validation of the proposed *VAAG MMH/FH Allocation Technique* has been carried out, for a typical fighter aircraft, assuming the generic systems. From the exercise, the MMH/FH goal for the systems/subsystems have been apportioned. However, this approach would be applicable in other domains where availability is of great essence and importance in achieving the design objective.

Acknowledgment. This work was executed in Aeronautical Development Agency, Bangalore. The authors would like to thank Mr. H Siddesha, Outstanding Scientist & Technology Director (LCA AF Mk2) for his support in executing this work.

References

1. Chipchak, J.S.: A practical method of maintainability allocation. IEEE Trans. Aerosp. Electron. Syst. **AES–7**(4), 585–589 (1971)
2. Fard, N.S.: Emanuel melachrinoudis maintenance scheduling for critical parts of aircraft. In: Proceedings Annual Reliability and Maintainability Symposium (1991)
3. Yanjie, Q., Zhigang, L., Bifeng, S.: New concept for aircraft maintenance management. In: Proceedings Annual Reliability and Maintainability Symposium, pp. 401–405 (2001)
4. Dupuy, M.J., Wesely, D.E., Jenkins, C.S.: Airline fleet maintenance trade-off analysis of alternate aircraft maintenance approaches. In: Proceedings of 2011 IEEE Systems and Information Engineering Design Symposium, April 29 (2011)
5. Zhang, A., Cui, L., Zhang, P.: Advanced military aircraft of study on condition-based maintenance. In: International Conference on Information Technology and Applications (2013)
6. Hlzel, N.B., Schilling, T., Gollnick, V.: An aircraft lifecycle approach for the cost-benefit analysis of prognostics and condition-based maintenance based on discrete-event simulation. In: Annual Conference of the Prognostics and Health Management Society (2014)

Determining the Balance Scorecard in Sheet Metal Industry Using the Intuitionistic Fuzzy Analytical Hierarchy Process with Fuzzy Delphi Method

S. Rajaprakash[1(✉)] and R. Ponnusamy[2]

[1] Department of Computer Science and Engineering, Sri Chandrasekarendra
Saraswathi Viswa Maha Vidyalaya [SCSVMV],
Enathur, Kanchipuram 631561, Tamil Nadu, India
srajaprakash_04@yahoo.com
[2] Department of Computer Science and Engineering,
Rajiv Gandhi College of Engineering, Chennai, India
r_ponnusamy@hotmail.com

Abstract. Balance Scorecard (BS) is an important part of human resource management in any organization or industry. It used to cascade the organization vision and its expectation and develop the employment capability. Balance scorecard may have many factors. In order to produce the best product and to retain the trust of customers, the industry should be able to identify which area has to be concentrated with higher priority in the Balance Scorecard. This situation lead with an uncertainty to multi criteria decision making. In this work, an attempt has been made for ranking the factors in the Balance Scorecard using Intuitionistic fuzzy analytical hierarchy process with fuzzy Delphi method.

Keywords: Intuitionistic fuzzy analytical hierarchy process · Analytical Hierarchy Process · Human resource

1 Introduction

A fuzzy set introduced by Zadeh (1965) defines a set characterized by a membership function. A membership function assigns to each element in the set under consideration a membership grade which is a value in the interval [0,1]. It was specifically designed to represent uncertainty and vagueness and to provide for normalized tools for dealing with the imprecision in intrinsic to many problems.

Fuzzy set introduces vagueness with the aim of reducing complexity by eliminating the sharp boundary dividing the members of the pair from

Currently Rajaprakash is research scholar at SCSVMV University and an Associate Professor at the Department of Computer Science and Engineering, Aarupadai Veedu Institute of Technology, Vinayaka Mission University Chennai, India.

© Springer International Publishing AG 2017
R. Prasath and A. Gelbukh (Eds.): MIKE 2016, LNAI 10089, pp. 105–118, 2017.
DOI: 10.1007/978-3-319-58130-9_11

non-members. This mapping associates each element in the set with its degree of membership. It can be expressed as a discrete value or as a continuous function. In fuzzy sets, each element is mapped by its membership function. They are triangular and trapezoidal membership functions that are commonly used for defining the continuous membership functions [1].

The triangular fuzzy membership function is given by

$$\mu_A(x) = \begin{cases} \frac{(x-a_1)}{(a_m-a_2)} : a_1 \le x \le a_m \\ \frac{(x-a_2)}{(a_m-a_2)} : a_m \le x \le a_2 \end{cases} \tag{1}$$

and the trapezoidal fuzzy membership function is given by

$$\mu_A(x) = \begin{cases} \frac{(x-a_1)}{(a_1^{(1)}-a_1)} : a_1 \le x \le a_1^{(1)} \\ 1 \qquad\quad : a_1^{(1)} \le x a_2^{(1)} \\ \frac{(x-a_2)}{a_2^{(1)}-a_2} : a_2^{(1)} \le x \le a_2 \end{cases} \tag{2}$$

1.1   Fuzzy Analytic Hierarchy Process (FAHP)

The Analytic Hierarchy Process (AHP) is one of the Fuzzy Multiple Criteria Decision Making methods. In 1983, Laahoven proposed the Fuzzy Analytical Hierarchy Process (FHAP). It is a combination of fuzzy set theory and Analytic Hierarchy Process. The application of FAHP is still flourishing. In this method, the ratio of the fuzzy comparison are able to better accommodate of vagueness comparing to AHP in which crisp value are used.

1.2   Intuitionistic Fuzzy Set (IFS)

Intuitionistic fuzzy set introduced by Atanassov [2]. The Intuitionistic fuzzy set theory is based on fuzzy set object and new objects and their properties. $0 \le \pi_A(x) \le 1$ for each $x \in X$ $\mu_A(x) \in [0,1]$ is the membership function of the fuzzy set $A^1 : \mu_{A^1}(x) \in [0,1]$ is the membership of $x \in A^1$.

The intuitionistic fuzzy set defined by

$$A = \{\langle x, \mu_x, \nu_x \rangle | x \in X\}, 0 \le \mu_x + \nu_x \le 1 \tag{3}$$

where $\mu_A : X \to [0,1]$ and $\nu_A : X \to [0,1]$ s.t $\mu_A(x) \in [0,1]$ denote the membership function and $\nu_A(x) \in [0,1]$ denote the non-membership function. obviously $A = \{\langle x, \mu_{A^1}(x), 1-\mu_{A^1}(x)\rangle | x \in X\}$ and $\pi_A(x) = 1-(\mu_x+\nu_x)$ is called the hesitation degree or degree of non-determinacy of $x \in A$ or x not $\in A$. Szmidt and Kacprzyk [3] point out that when calculating the distance between two IFSs, we can not omit $\pi_A(x)$. We consider that $\alpha = (\mu_\alpha, \nu_\alpha, \pi_\alpha)$ is an intuitionistic fuzzy values where $\mu_\alpha \in [0,1]$ and $\nu_\alpha \in [0,1], \mu_\alpha + \nu_\alpha \le 1$. According to the Szmidt and Kacprzyk [3] put forth a function in mathematical form

$$\rho(\alpha) = 0.5(1 + \pi_\alpha)(1 + \mu_\alpha) \tag{4}$$

The α means that it contains all positive information included. Therefore intuitionistic fuzzy set is mainly based on the membership function, the non-membership function, and the indeterminacy degree.

Table 1. Comparison scale [4]

Linguistic value	Scale	Linguistic scale
9	0.9	Extreme important
7	0.8	Very strong important
5	0.7	Strong important
3	0.6	Moderately important
1	0.5	Equal preference
1/3	0.4	Moderately not important
1/5	0.3	Strong not important
1/7	0.2	Very strong not important
1/9	0.1	Extreme not important

1.3 Intuitionistic Relation

Let R be the relation in the intuitionistic values on the set $X = \{x_1, x_2...x_n\}$ is represented by matrix $R = (M_i^k)_{n \times n}$, where $M_{ik} = \langle (x_i, x_k), \mu(x_i, x_k), \nu(x_i, x_k) \rangle$ $i, k = 1, 2, 3, \cdots, n$. Let us assume that $M_{ik} = (\mu_{ik}, \nu_{ik})$ and $\pi(x_i, x_k) = 1 - \mu(x_i, x_k) - \nu(x_i, x_k)$ is interpreted as an indeterminacy degree. The notion of intuitionistic fuzzy $t - norm$ and $t - conorm$ is as found in Deschrijver et al. [5]. The intuitionistic fuzzy triangular norms was studied by Xu [4]. He introduced the following operations

1. $M_{ik} \bigoplus M_{lm} = (\mu_{ik} + \mu_{lm} - \mu_{ik}\mu_{lm}, \nu_{ik}\nu_{lm})$
2. $M_{ik} \bigotimes M_{lm} = (\mu_{ik}\mu_{lm}, \mu_{ik} + \mu_{lm} - \nu_{ik}\nu_{lm})$

In our work, we are applying the Intuitionistic Fuzzy AHP with Delphi method, over the balance scorecard in Auto Sheet Metal Industry, India. Based on the above scale, Table 1 shows the comparison matrix with the ranked list of factors.

1.4 Fuzzy Delphi Method

Kaufman and Gupta [6] have studied about the Fuzzy Delphi Method. In 1993, Ishikawa et al. [7] also mention about the Fuzzy Delphi Method. The generalization of fuzzy Delphi method is as follows:

1. Identify the experts based on the domain and make the experts panel members
2. Based on the experts opinion identify the attributes and categories. Using the attributes make the questionnaires.
3. Using the questionnaires gets the first round of the suggestion about the attributes.
4. From the attributes values compute the Mean. [8] Then deviation is calculated between mean and each expert opinion. [it is also a fuzzy number]. The deviation is sent to each expert for revaluation.

5. In the second round a new fuzzy number is received from the experts. The same procedure is repeated (step-2) until two successive means become very close; else the Delphi expert will take the final decision.

2   Literary Survey

Satty [9] introduced the AHP approach for decision making. Atanassov [2] proposed the intuitionistic fuzzy sets and its applications. Mahman Akkram et al. [10] proposed an approach to control the heat produced by fans using the Intuitionistic fuzzy logic. In this work, the heat of the fan is calculated with the help of intuitionistic fuzzy rules applied in an inference engine using defuzzification method. The Intuitionistic fuzzy sets are used in some medical application by Eulalia Szmidt et al. [11]. As a generalization of fuzzy sets, a new definition of distance between two intuitionistic fuzzy sets has been suggested by Atanassov et al. [3].

Rehan Sadiq et al. [12] proposed the intuitionistic fuzzy analytic hierarchy process to make the environmental decision making process easier. In this work, authors find the best drilling operation in the fluid(mud). Determining the customer satisfaction in the automobile sector, the Intuitionistic fuzzy analytic hierarchy process has been studied by Rajaprakash et al. [13]. To increase the hotel atmospheres and its environment, use of Delphi fuzzy Analytical Hierarchy Process ranking the customer preference in spa atmosphere has been studied in two phases: the first one is Delphi method and the second one is by AHP, by Yen Cheng Chen et al. [14]. The selection of the best DBMS among the several candidates in the Turkish National Identity Card Management project was done using the Fuzzy AHP by F. Ozgur Catak et al. [15]. Using the Fuzzy AHP evaluation of the E-commerce in order to manage and determine the drawbacks, opportunities were studied by Feng Kong et al. [21]. The supplier selection is typically a multi criteria decision problem. Ranking the similarity to ideal solution (TOPSIS) method for group decision making with intuitionistic fuzzy numbers is proposed by Mohammad Izadikhah [16]. The prediction of the highest and the lowest temperature by BP neural networks training for abnormal weather alerts have been studied by Dan Wang et al. [8] using a fuzzy AHP and rough set. In this work, authors compared the fuzzy AHP and rough set.

The expectation of students in the present education system in Tamilnadu, India has been studied using the FAHP. In the work, Rajaprakash et al. [17] have taken a sample work on Engineering education. The Fuzzy Delphi Method and Fuzzy analytic Hierarchy process have been applied to determine the critical factors of the regenerative technologies and find the degree of each importance criterion as the measurable indices of the regenerative technologies. This work was attempted by Yu-Lung Hsu et al. [7]. The study of human capital indicator and ranking by using IFAHP to evaluate the four main indicators of Human capital has been studied by Lazim Abdullah et al. [18]. Diagnosis progress in bacillus colonies identification in the medical field using the intuitionistic fuzzy set theory studied by Hoda davarzani et al. [18]. Intuitionistic Fuzzy Delphi

Method used as forecasting tool based on the suggestion of the experts in the work proposed by Tapan Kumar *et* al. [19]. They used triangular fuzzy number and aggregation process based on the opinion of the experts.

3 Methodology

1. Using the fuzzy Delphi method experts, values are converted to Intuitionistic value and make the comparison matrix (comparison scale Table 4).
2. To the check the consistent of the matrix intuitionistic preference relation calculated based on Xu *et* al. [20] which is given below
$R = (M_{ik})_{n \times n}$ with $M_{ik} = (\mu_{ik}, \nu ik)$ is multiplicative consistent if

$$\mu_{ik} = \begin{cases} 0 & if (\mu_{it}, \mu_{tk}) \in \{(0,1),(1,0)\} \\ \frac{\mu_{it}\mu_{tk}}{\mu_{it}+\mu_{tk}+(1-\mu_{it})(1-\mu_{tk})} & otherwise \end{cases} \tag{5}$$

$$\nu_{ik} = \begin{cases} 0 & if (\nu_{it}, \nu_{tk}) \in \{(0,1),(1,0)\} \\ \frac{\nu_{it}\nu_{tk}}{\nu_{it}+\nu_{tk}+(1-\nu_{it})(1-\nu_{tk})} & otherwise \end{cases} \tag{6}$$

Theorem [20]: In fuzzy preference relation, the following statement are equivalent:

$$b_{ik} = \frac{b_{ik}b_{tk}}{b_{ik}b_{tk} + (1 - b_{ik})(1 - b_{tk})} \quad i, t, k = 1, 2, 3... \tag{7}$$

$$b_{ik} = \frac{\sqrt[n]{\prod_{s=1}^{n} b_{ik}b_{tk}}}{\sqrt[n]{\prod_{s=1}^{n} b_{is}b_{sk}} + \sqrt[n]{\prod_{s=1}^{n} b_{is}b_{sk}}} \quad i, k = 1, 2, ...n \tag{8}$$

$$\bar{\mu}_{ik} = \frac{\sqrt[k-i-1]{\prod_{t=i+1}^{k-1} \mu_{it}\mu_{tk}}}{\sqrt[k-i-1]{\prod_{t=i+1}^{k-1} \mu_{it}\mu_{tk}}} \quad k > i+1 \tag{9}$$

$$\bar{\nu}_{ik} = \frac{\sqrt[k-i-1]{\prod_{t=i+1}^{k-1} \nu_{it}\nu_{tk}}}{\sqrt[k-i-1]{\prod_{t=i+1}^{k-1} \nu_{it}\nu_{tk}}} \quad k > i+1 \tag{10}$$

3. The distance between intuitionistic relation [3] is calculated using

$$d(M, \bar{M}) = \frac{1}{2(n-1)(n-2)} \sum_{t=1}^{n}\sum_{k=1}^{n} (|\bar{\mu}_{ik} - \mu_{ik}| + |\bar{\nu}_{ik} - \nu_{ik}| + |\bar{\pi}_{ik} - \pi_{ik}|) \tag{11}$$

4. The priority of the intuitionistic preference relation zeshuri Xu [20] is calculated by the following method.

$$W_i = \frac{\sum_{k=1}^{n} M_{ik}^1}{\sum_{i=1}^{n} \sum_{k=1}^{n} M_{ik}^1}$$

$$W_i = \left[\frac{\sum_{k=1}^{n} \mu_{ik}}{\sum_{i=1}^{n} \sum_{k=1}^{n} [1 - \nu_{ik}]}, 1 - \frac{\sum_{k=1}^{n} [1 - \nu_{ik}]}{\sum_{i=1}^{n} \sum_{k=1}^{n} \mu_{ik}} \right] \tag{12}$$

5. After finding the weights of the all levels based on weights (ranking the weight) by using the Eq. 4 then finding preference ranking.

4 Illustrative Work

The above work illustrated in area of the balance scorecard of human resource in the automobile sector sheet metal industry at Chennai. Balanced Scorecard is classified into various factors based on the experts opinion. Here we are ranking the factor of Balanced Scorecard using the above method. In this work, the data collected from sheet metal industry.

4.1 Balance Scorecard

The balanced scorecard revolutionized conventional thinking about performance metrics. In 1992 Kaplan and Norton first introduced the concept. The scorecard allowed companies to track financial results while monitoring progress in building the capabilities needed for growth. It is used to strengthen the Key Resource Area's (KRA) and Key performance Indicator's (KPI'S) of every individual which helps for higher productivity, cost efficiency, effective utilization of machine and man and method. Balance scorecard are categorised into four important attributes like Financial, Customer, Internal Process, Learning and growth in level-1. In level-2, each attributes divided into sub attributes.

4.2 Finance

Based on the observation from the experts the Balance Score Card is used to promote the business and its future requirements via formulating the critical needs of the business with meticulously framed Metrics to cater their clients in the most efficient way as well as coping up with the latest technologies. The important factor are like Profit, Growth, People and Reputation. The process is adopted half yearly for effective monitoring and implementation.

How effective the people are utilized. How effective the assets fixed and variable has been utilized and handled. If all the above points have been used then organization achieves the best revenue in business.

4.2.1 Achieve Total Revenue of Business
1. It is based on the following queries.
2. What should be the sales number of the current year?
3. What should be the manpower head count bared on the sales?

4. What should be the welfare expenses bared on the sales as well manpower head count?
5. What are the effective ways to control the cost?
6. What kind of economic cost effective automatons has to be brought in?
7. How much number of Kaizens has to be evolved?
8. How well the plant maintenance has to be adopted?

4.2.2 Collecting the Overdue Debits

The finance department of the organization has to separate the good and bad vendors. Accordingly the debits can also been categorized as immediate receivables with regulars intervals, recovered only after regular follow up, recovered when only any employees goes on personal visits. In short, the organization has to fix its price according to their rate of return to overcome the loss in overdue debits.

4.2.3 Achieve the Targeted Rate of Return

To achieve the targeted rate of return, the organization should focus on the achieving the total revenue of the business, collecting the overdue debits, as well targeting on its employee productivity. In short, it has to booze up the Man Method and the machines.

4.2.4 Inventory Turnover

The organization has to concentrate on its inventory to have a control on it. It is also helps in huge cost reduction, the success of maintain inventory lies in how many times the inventory has been turnover, and how much cost has been saved. The art of managing inventory lies in keeping the minimal stock by avoiding ageing of inventory and efficiently managing by effective utilization of the factory. Generally the inventory turnover formula is average is stock and number of days in month divided by sales.

The main support for inventory monitoring is the stores department. The store will depict the assessment of inventories. If the stores have been monitored well with the options of labeling, arranging a proper accessing enables the success of efficient inventory turnover.

4.3 Customer

The customer is the important role in the sheet metal industry, here customer satisfaction and customer requirement are very important part in Business Scorecard. When the customer relationship is honored and the industry can get a new Business, which reflects the overall branding of the industry. Customer relationship include many factors like quality, delivery, commercial dealing, product quality, response to queries, concern of problem etc. So based on the many factors customer relationship honoured Rajaprakash *et* al. [13]. The customer requirement is based on the many factor like Quality, Cost,

Delivery, and service, Supplier request for Engineering change approval(SREA), Emergency Response action(ERA), Corrective action and Preventive action taken(CA and PA) and so on, Similarly Based on the customer requirement and satisfaction a matrix has been framed it is known as customer matrix. Thus from the customer matrix we know the clear picture of the customer. If the customer satisfaction is honoured then the industry can get the new business.

4.4   Internal Process

The internal process is the main back none for the success of the organization business. The internal process is decided by improved material, Improve VAPE and PPM set by customer plant maintenance, Tier II development and 2s safety.

4.4.1   Improve Material

The material consumption has to be followed properly and rigorously for the success of the business. The material planning is the vital role of the organization. The material planning team has to keep constant touch with the production planning team to anticipate the production fluctuating and ramp up. This coordination will help to keep the material stock in control as well to produce the emergency material which can create a crisis situation in the shop floor. This kind of planning always helps the last hour rushing as well avoiding freight cost which normally occurs and business the EBITA.

4.4.2   Improve VAPE

The value addition per employee is the main factor or calculative method which helps to calculate and control the employee recruitment cost and their effectiveness monitoring in the organization in the sense how the new employee or the exiting employee contributes in regards with their cost to company the VAPE is calculated as follows. VAPE is equal to value added divided by number of employees. It is decided and directly proportional to the VAPCO. The VAPCO is derived with value addition per employee cost. VAPCO is equal to value added per manpower cost. The value added is equal to sales − (raw material cost + Job work + VAPE). Manpower cost equal to regular + contract worker + staff welfare. It is basically used to calculate the head count of the employees with its sales and to benchmark the best of the head count required. It will be further fine tuned by required continuous manpower optimization with proper job description, work allocation, work planning and so on.

4.4.3   PPM Set by Customer

The PPM will be set by customer and it has to be followed in reward with quality and target has to be fixed for it. The PPM has to continuously monitored by internal and external quality circle.

4.4.4   Plant Maintenance

The plant has to be maintained in its regular manner. Generally the idea is production industry is the shop floor depicts the organization technology values and principles. The plant is well maintained by various initiates like 5s, TPM, etc. The well maintained plant always reflects high productivity and high morale.

4.4.5   2s and Safety

The 2s safety has to be maintained strictly since there in no compromise on safety otherwise it is zero tolerance with safety. The 2s helps the operator to finish his task sap as well the tools will be fixed and available in the required slot. It will reduce the time loss and save the energy of the operator.

4.5   Learning and Growth

The HR Department should work for the development of their people and welfare through conducting various Employee Engagement Activities. The organization should identify the potential of the business and manage the latest technology and competition thereby, bringing in competitive products, people and price as well contributing to the brand building of the organization Fig. 1.

4.6   Business Scorecard in Level-1

In order to find the Balance Scorecard in Level-1 four attributes are available. Based on the experts opinion the first initial Table 2 formed

Table 2. Delphi 1

Experts	BS1 to BS2		BS1 to BS3		BS1 to BS4		BS2 to BS3		BS2 to BS4		BS3 to BS4	
1	0.5	0.4	0.4	0.2	0.6	0.3	0.4	0.3	0.5	0.3	0.6	0.4
2	0.6	0.3	0.5	0.3	0.5	0.4	0.3	0.4	0.7	0.3	0.6	0.4
3	0.7	0.3	0.3	0.2	0.5	0.3	0.5	0.3	0.5	0.5	0.5	0.6
4	0.3	0.6	0.6	0.3	0.6	0.3	0.4	0.4	0.5	0.4	0.7	0.4
5	0.4	0.5	0.2	0.3	0.3	0.6	0.5	0.5	0.6	0.4	0.5	0.4
6	0.5	0.4	0.3	0.4	0.5	0.5	0.6	0.4	0.3	0.5	0.6	0.5
7	0.6	0.3	0.4	0.2	0.5	0.4	0.7	0.3	0.5	0.4	0.5	0.6
8	0.7	0.2	0.3	0.3	0.6	0.5	0.3	0.4	0.6	0.4	0.6	0.4
9	0.5	0.4	0.2	0.5	0.4	0.4	0.5	0.5	0.7	0.4	0.4	0.5
10	0.4	0.5	0.7	0.3	0.3	0.6	0.5	0.2	0.6	0.4	0.5	0.6

The Mean values are calculated. The deviations of experts opinion from the calculated Mean values are given below Table 3.

Here the Delphi expects not satisfied with deviation Table 3. Therefore the opinion is sent back to the experts for one more opinion.

Now the Delphi expert is satisfied with the above deviation Table 5. Based on the expert suggestion the first intuitionistic preference relation matrix BS formed is shown below.

$$M = \begin{pmatrix} (0.5, 0.5) & (0.5, 0.4) & (0.4, 0.6) & (0.5, 0.6) \\ (0.7, 0.3) & (0.5, 0.5) & (0.5, 0.4) & (0.5, 0.5) \\ (0.5, 0.4) & (0.5, 0.4) & (0.5, 0.5) & (0.4, 0.5) \\ (0.6, 0.4) & (0.6, 0.4) & (0.5, 0.4) & (0.5, 0.5) \end{pmatrix}$$

The deviation from the mean is calculated Table 5.

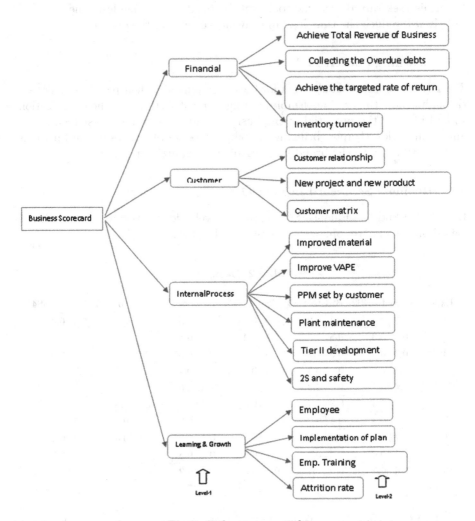

Fig. 1. Balance scorecard

Table 3. Delphi2

Experts	BS1 to BS2		BS1 to BS3		BS1 to BS4		BS2 to BS3		BS2 to BS4		BS3 to BS4	
1	0.02	−0.01	−0.01	0.1	−0.12	0.13	0.07	0.07	0.05	0.1	−0.05	0.08
2	−0.08	0.09	−0.11	0	−0.02	0.03	0.17	−0.03	−0.15	0.1	−0.05	0.08
3	−0.18	0.09	0.09	0.1	−0.02	0.13	−0.03	0.07	0.05	−0.1	0.05	−0.12
4	0.22	−0.21	−0.21	0	−0.12	0.13	0.07	−0.03	0.05	0	−0.15	0.08
5	0.12	−0.11	0.19	0	0.18	−0.17	−0.03	−0.13	−0.05	0	0.05	0.08
6	0.02	−0.01	0.09	−0.1	−0.02	−0.07	−0.13	−0.03	0.25	−0.1	−0.05	−0.02
7	−0.08	0.09	−0.01	0.1	−0.02	0.03	−0.23	0.07	0.05	0	0.05	−0.12
8	−0.18	0.19	0.09	0	−0.12	−0.07	0.17	−0.03	−0.05	0	−0.05	0.08
9	0.12	−0.11	−0.31	0	0.18	−0.17	−0.03	0.17	−0.05	0	0.05	−0.12
10	0.12	−0.11	−0.31	0	0.18	−0.17	−0.03	0.17	−0.05	0	0.05	−0.12

Table 4. Delphi3

Experts	BS1 to BS2		BS1 to BS3		BS1 to BS4		BS2 to BS3		BS2 to BS4		BS3 to BS4	
1	0.6	0.4	0.4	0.4	0.6	0.3	0.4	0.4	0.52	0.5	0.3	0.5
2	0.6	0.4	0.3	0.5	0.6	0.4	0.4	0.4	0.6	0.6	0.4	0.4
3	0.5	0.4	0.6	0.2	0.6	0.2	0.5	0.4	0.4	0.4	0.4	0.6
4	0.6	0.6	0.4	0.4	0.6	0.3	0.4	0.4	0.6	0.4	0.4	0.6
5	0.6	0.4	0.5	0.4	0.5	0.6	0.5	0.5	0.5	0.5	0.3	0.6
6	0.4	0.2	0.4	0.4	0.5	0.5	0.6	0.4	0.6	0.5	0.5	0.4
7	0.6	0.4	0.4	0.4	0.5	0.4	0.7	0.3	0.5	0.5	0.4	0.3
8	0.4	0.4	0.5	0.5	0.4	0.4	0.4	0.4	0.5	0.5	0.5	0.5
9	0.4	0.4	0.3	0.4	0.4	0.4	0.6	0.4	0.4	0.5	0.4	0.6
10	0.5	0.5	0.4	0.3	0.4	0.5	0.5	0.4	0.4	0.6	0.5	0.4

Table 5. Delphi4

Experts	BS1 to BS2		BS1 to BS3		BS1 to BS4		BS2 to BS3		BS2 to BS4		BS3 to BS4	
1	−0.08	0.01	0.02	−0.01	−0.09	0.1	0.1	0	−0.01	0 0.11	−0.01	0
2	−0.08	0.01	0.12	−0.11	−0.09	0	0.1	0	−0.098	−0.1	0.01	0.09
3	0.02	0.01	−0.18	0.19	−0.09	0.2	0	0	0.102	0.1	0.01	−0.11
4	−0.08	−0.19	0.02	−0.01	−0.09	0.1	0.1	0	−0.098	0.1	0.01	−0.11
5	−0.08	0.01	−0.08	−0.01	0.01	−0.2	0	−0.1	0.002	0	0.11	−0.11
6	0.12	0.21	0.02	−0.01	0.01	−0.1	−0.1	0	−0.098	0	−0.09	0.09
7	−0.08	0.01	0.02	−0.01	0.01	0	−0.2	0.1	0.002	0	0.01	0.19
80	.12	0.01	−0.08	−0.11	0.11	0	0.1	0	0.002	0	−0.09	−0.01
90	.12	0.01	0.12	−0.01	0.11	0	−0.1	0	0.102	0	0.01	−0.11
10	0.02	−0.09	0.02	0.09	0.11	−0.1	0	0	0.102	−0.1	−0.09	0.09

To check the consistence preference relation using the above formula (9) and (10) we can get the multiplicative fuzzy relation Matrix($\bar{M}$).

$$\bar{M} = \begin{pmatrix} (0.5, 0.5) & (0.5, 0.4) & (0.4494, 0.5) & (0.4, 0.51) \\ (0.6, 0.5) & (0.5, 0.5) & (0.5, 0.4) & (0.5025, 0.4449) \\ (0.5, 0.449) & (0.4, 0.5) & (0.5, 0.5) & (0.4, 0.5) \\ (0.5, 0.4) & (0.449, 0.50254) & (0.5, 0.4) & (0.5, 0.5) \end{pmatrix}$$

Then the Eq. 11 calculates, the distance between intuitionistic relation is calculated as $d(\bar{M}, M) = 0.09578$ which is less than τ. Here we will fix the threshold value as $\tau = 0.1$. Therefore the above matrix is consistent. The next step is calculating the weight of all attributes using the Eq. 12. It is given in the Table 6 and using the Eq. 4, we will get the preference(P) of all the attributes. It is given in the Table 7. We can see that the first preference is finance, second is customer, third is internal process and last one is Learning growth. Similarly we calculated the other levels.

Table 6. Weight

weight	μ	ν
W(BS1)	0.21763	0.72733
W(BS2)	0.2474	0.702077
W(BS3)	0.2118	0.73375
W(BS4)	0.22939	0.71467

Table 7. Preference

Attribute	$\rho(\alpha)$	P
Finance(BS1)	0.88958	1
Customer(BS2))	0.85299	2
Internal process(BS3)	0.852624	3
Learning and growth(BS4)	0.840044	4

5   Comparison Study

Comparing the work IFAHP with Delphi system with IFAHP, We can see that our work has so many advantages over IFAHP method.

1. In the work of IFAHP proposed by Zeshui Xu [4], he used two algorithms. In Algorithm-I, the deviation is calculated using Eq. 11. Suppose the deviation is too high then the author uses Algorithm-II. If still there is no consistency, he starts all over again.
   In the present work we use algorithm-I of Xu [4] and calculate the deviation. The deviation is mostly less than the threshold value $\rho(\alpha)$ since we have done enough iterations in the fuzzy Delphi method.
2. In Zeshuri Xu [4], the work has not given clear picture of how to go about it when Algorithm-II fails. In our present work, we highlighted the following (in step-4): if the preference relation is inconsistent then, go to step-1 (Fuzzy Delphi Method).
3. In Zeshuri Xu [4] work, ranking the supplier in the last level (that is in the alternative criteria only). But in our work, we calculated the preference of each level so that it useful at each and every level of the work.

6   Empirical Result

Based on the suggestion given by the experts the Business Scorecard hierarchy is formed in the Level-1. In that four attributes(factors), the first preference is finance, second is customer, third one is internal process and the last preference is Learning and Growth Table 7. Similarly in the level-2, Finance have four sub attributes after calculating using IFAHP with fuzzy Delphi method, the first preference is achieve total revenue of business, second is achieve the targeted rate of return, third preference is inventory turnover and the last preference is collecting the overdue debts. In Level-2, Customer have three attributes and its ranked based on the above method New product and project, customer relationship and Customer matrix respectively. In the Level-2 one the main attributes is internal process have six sub attributes in the first preference is improve VAPE, second is PPM set by customer, third preference is improved material, fourth one is plant maintenance, fifth is Tire II development and last preference is 2s safety.

7   Conclusion

Recently many of researches are interested to apply the Intuitionistic Fuzzy set in their problems. This work combines Intiutionistic Fuzzy Analytical Hierarchy Process and Fuzzy Delphi Method over the Business Scorecard in the automobile sector in India. The Major part of IFAHP With Fuzzy Delphi Method include the following: In Delphi Method questionnaires are framed based on the suggestion and get the opinion from the experts in automobile sector. This survey was taken using the web site Surveymonkey.com (here the questions are framed and sent link to the experts). In this work we are categories the Business Scorecard. In Each and every level, we rank (preference) the factors of Business Scorecard. The major disadvantage of our work is identify the experts and getting opinion from the experts will take lot of time. Which is useful for the automobile sector and it may improve the our industrial standard and Indian economy.

References

1. Klir, G.J.: Fuzzy Set and Fuzzy Logic Theory and Application. PTR Publisher, New York (1995)
2. Atanassov, K.T.: Intuitionistic fuzzy sets. Fuzzy Sets Syst. **20**(1), 87–96 (1986)
3. Szmidt, E., Kacprzyk, J.: Distances between intuitionistic fuzzy sets. Fuzzy Sets Syst. **114**(3), 505–518 (2000)
4. Xu, Z., Liao, H.: Intuitionistic fuzzy analytic hierarchy process. IEEE Trans. Fuzzy Syst. **22**(4), 749–761 (2014)
5. Deschrijver, G., Cornelis, C., Kerre, E.E.: On the representation of intuitionistic fuzzy t-norms and t-conorms. IEEE Trans. Fuzzy Syst. **12**(1), 45–61 (2004)
6. Kaufmann, A., Gupta, M.M.: Fuzzy Mathematical Models in Engineering and Management Science. Elsevier Science Inc., New York (1988)

7. Hsu, Y.L., Lee, C.H., Kreng, V.B.: The application of fuzzy delphi method and fuzzy ahp in lubricant regenerative technology selection. Expert Syst. Appl. **37**(1), 419–425 (2010)
8. Carlsson, C., Fullér, R.: On possibilistic mean value and variance of fuzzy numbers. Fuzzy Sets Syst. **122**(2), 315–326 (2001)
9. Saaty, T.: The Analytic Hierarchy Process, Planning, Priority Setting, Resource Allocation. McGraw-Hill, New York (1980)
10. Akram, M., Shahzad, S., Butt, A., Khaliq, A.: Intuitionistic fuzzy logic control for heater fans. Math. Comput. Sci. **7**(3), 367–378 (2013)
11. Szmidt, E., Kacprzyk, J.: Intuitionistic fuzzy sets in some medical applications. In: Reusch, B. (ed.) Fuzzy Days 2001. LNCS, vol. 2206, pp. 148–151. Springer, Heidelberg (2001). doi:10.1007/3-540-45493-4_19
12. Sadiq, R., Tesfamariam, S.: Environmental decision-making under uncertainty using intuitionistic fuzzy analytic hierarchy process (IF-AHP). Stoch. Env. Res. Risk Assess. **23**, 75–91 (2009)
13. Rajaprakash, S., Ponnusamy, R., Pandurangan, J.: Determining the customer satisfaction in automobile sector using the intuitionistic fuzzy analytical hierarchy process. In: Prasath, R., O'Reilly, P., Kathirvalavakumar, T. (eds.) MIKE 2014. LNCS (LNAI), vol. 8891, pp. 239–255. Springer, Cham (2014). doi:10.1007/978-3-319-13817-6_24
14. Chen, Y.C., Yu, T.H., Tsui, P.L., Lee, C.S.: A fuzzy ahp approach to construct international hotel spa atmosphere evaluation model. Quality **48**(2), 645–657 (2014)
15. Catak, F.O., Karabas, S., Yildirim, S.: Fuzzy analytic hierarchy based DBMS selection in Turkish National Identity Card Management project. Int. J. Inf. Sci. Tech. (IJIST) **2**(4), 29–38 (2012)
16. Izadikhah, M.: Group decision making process for supplier selection with TOPSIS method under interval-valued intuitionistic fuzzy numbers. Adv. Fuzzy Syst. **2012**(2), 2 (2012)
17. Rajaprakash, S., Ponnusamy, R.: Determining students expectation in present education system using fuzzy analytic hierarchy process. In: Prasath, R., Kathirvalavakumar, T. (eds.) MIKE 2013. LNCS (LNAI), vol. 8284, pp. 553–566. Springer, Cham (2013). doi:10.1007/978-3-319-03844-5_55
18. Abdullah, L., Jaafar, S., Taib, I.: Intuitionistic fuzzy analytic hierarchy process approach in ranking of human capital indicators. J. Appl. Sci. **13**(3), 423–429 (2013)
19. Tapan Kumar, R., Garai, A.: Intuitionistic fuzzy delphi method: more realistic and interactive forecasting tool. Notes Intuitionistic Fuzzy Sets **18**(50), 37–50 (2012)
20. Xu, Z.: Intuitionistic preference relations and their application in group decision making. Inf. Sci. **177**(11), 2363–2379 (2007)
21. Kong, F., Liu, H.: Appling fuzzy analytic hierarchy process to Evaluate Success Factors of E-Commerce. Int. J. Inf. Syst. Sci. **1**(3–4), 406–412 (2005)

Identifying and Pruning Features for Classifying Translated and Post-edited Gaze Durations

Tanik Saikh, Dipankar Das, and Sivaji Bandyopadhayay[(⊠)]

Department of Computer Science and Engineering,
Jadavpur University, Kolkata, India
tanik4u@gmail.com, dipankar.dipnil2005@gmail.com,
sivaji_cse_ju@yahoo.co.in

Abstract. The present paper reports on various experiments carried out to classify the source and target gaze fixation durations on an eye tracking dataset, namely Translation Process Research (TPR). Different features were extracted from both the source and target parts of the TPR dataset, separately and different models were developed separately by employing such features using a machine learning framework. These models were trained using Support Vector Machine (SVM) and the best accuracy of 49.01% and 59.78% were obtained with respect to cross validation for source and target gaze fixation durations, respectively. The experiments were also carried out on the post edited data set using same experimental set up and the highest accuracy of 71.70% was obtained. Finally, Information Gain based pruning has been performed in order to select the best features that are useful for classifying the gaze durations.

Keywords: Eye tracking · Gaze fixation duration · Translation · Post-editing · Information gain

1 Introduction

The term, '*Gaze*' has been defined differently in different studies. In Psychology, gaze duration refers to the sum of all fixations of a particular word before the eyes move to a different word [5]. In translation studies, gaze duration is often used to describe as total reading time, i.e. the sum of all fixations of a word, irrespective of its occurrences [6–9]. In general, during fixation, the gaze remains on a single location for the duration of several milliseconds. If the gaze is observed in the source window, it is called as source gaze and defined as '*TrtS*' in the dataset whereas it is termed as target gaze '*TrtT*' if the gaze is observed in the target window [11, 12]. The gazes are measured in milliseconds say for example 0 ms, 180 ms etc.

With the advent of social media and rapid growth of code mixed data, a new version of language is being evolved and this new version is difficult to be processed because we mix our native languages with several other languages [17]. However, it is also crucial and even challenging if we want to translate such native texts into their equivalent texts and vice versa. Thus, in this era of globalization and mobilization,

© Springer International Publishing AG 2017
R. Prasath and A. Gelbukh (Eds.): MIKE 2016, LNAI 10089, pp. 119–131, 2017.
DOI: 10.1007/978-3-319-58130-9_12

people find themselves in a platform where translation issues are very much involved. Center for Research and Innovation in Translation and Translation Technology (CRITT) at Copenhagen Business School, basically focus on the research and development of new knowledge regarding translation and communication processes for last 10 years. Over the years, CRITT is carrying out innovative researches and addressing these issues by conducting various summer projects and workshops. They collected the Translation Process Research (TPR) dataset by using Translog-II tool[1] [1, 2] and the rudimentary version was released in 2012 as Translation Process Research Database (TPR-DB) [3, 4]. Later, many more researches have been carried out and several features were added to TPR-DB. A huge amount of machine translated and post edited TPR data was collected as a part of the "Cognitive Analysis and Statistical Methods for Advanced Computer Aided Translation" (CASMACAT)[2] project [13] and made it publicly available in creative common license. TRP-DB data basically contains User Activity Data (UAD) of translator's behavior such as gaze, translator's activities etc.

In literature, many tasks are found on eye gaze in psycholinguistic field [18, 19]. To the best of our knowledge, no such significant work has been carried out on gaze prediction or classification in translation studies except a few described in [14] where a machine learning approach was adapted to predict the source gaze fixation duration. They extracted several features and revealed the importance of those features in identifying gaze fixation duration.

This research topic has many application areas, e.g., if one is capable to predict gaze behavior, of course s/he is also capable to deeply look into the mental state of that person. That will (fortunately!) never be 100% possible. However, it is bit imaginable to identify the quality of a reader as 'good' or 'bad' by predicting their gazes. Also for the Governments, it might be interesting to better control/survey their people by showing up explanations/definitions where s/he gets stuck and diverting to something else if boring etc. All this is likely to be exploited eventually by companies to better sell their stuff, and to brainwash us even more than they are already doing.

The task reported in this paper aims to predict the gazes of source and target tokens using a machine learning framework based on a rigorous analysis of various features. At first, the gaze to word mapping data was collected from the summary table along with various information of translation sessions like composed and basic product units, composed and basic process units etc. The source token (ST) files from the basic product units have been considered to accomplish our task. The experiments have been carried out on the TPR-DB dataset (version 2.1) that contains word to gaze mapped instances along with various features. The instances are grouped into four classes according to their respective gaze durations. We extracted different features from the dataset and various models are built by combining different features. The classification models are trained using a supervised learning method, namely Support Vector Machine. Different classification accuracies have been achieved for various models. We achieved 49.01% and 59.78% classification accuracies for classifying the source and target gaze fixation durations, respectively. The experiments on feature analysis

---

[1] https://sites.google.com/site/centretranslationinnovation/translog-ii.

[2] http://www.casmacat.eu/.

have been conducted on two sets, one is on translation data containing both the source and target and other is only on the post-edited part. Some of the additional experiments performed are enlisted below.

(1) Experiments on translation data (KTHJ08 study):
    a. A usual feature analysis has been followed by an ablation study that includes the Pearson's correlation coefficient [16] with respect to every feature towards our predicting value.

(2) Experiments on post editing data (BML12 study):
    a. Here, instead of using all the features, only 4 features (namely Translator's Identity, Unigram and Bigram probability, Parts of speech) have been used. Combining these features, several models are built and trained using a supervised learning approach.
    b. Deeper feature analysis has been performed for further classifying the instances into their core-grained classes in contrast to their initial assigned classes.

(3) Information Gain based pruning has also been carried out for selecting the important features that are useful for classifying the source-target and post-edited gaze instances.

2 Data

One of the basic aims for conducting any multilingual experiment is to compare the performances of different translators from the perspectives of *translation* (T), *post-editing* (P), and *monolingual post-editing* (E) for different languages from scratch. A total of six source texts in English are translated by students who are experienced translators. Three texts (1–3) are taken from news whereas the rest three texts (4–5) are collected from sociological texts of encyclopedia. This database is publicly available[3] and contains recorded translation sessions. It also contains User Activity Data (UAD) of translator's behaviors, collected from almost 30 studies of translation, post-editing, revision, authoring and copying tasks and recorded with CASMACAT workbench and Translog-II. In the present attempt, only the database KTHJ08 has been used for our experiment and this study contains only translation data for news text (1–3). The study of KTHJ08 contains English to Danish translation data. Only 69 source text files have been considered along with a total of 10640 instances.

3 Feature Analysis

Feature selection is a pivotal part of any machine learning based experiment in order to improve the performance of a system. Therefore, in our experiment, only those features which play the important roles for identifying our predicting variable are discussed in

---

[3] https://sites.google.com/site/centretranslationinnovation/tpr-db.

this section. The descriptions of such features present in the dataset are as follows. We also fit a linear regression model available with Weka toolkit for analyzing these features against our predicting values and it shows better correlation coefficients of those features. The following Table 1 depicts some features and it's corresponding values.

Table 1. Features and it's corresponding correlation coefficient

Feature: Pearson correlation coefficient
Part: 0.2958, **prob1:** 0.2317, **prob2:** 0.1201, **Pos:** 0.3219, **Length:** 0.3784, **SAUnbr:** −0.0242, **InEff:** −0.0242, **Cross:** −0.0242, **Munit:** 0.0467, **Htra:** 0.173

3.1 Part (*Part*)

This feature denotes basically the number of different participants. In case of our dataset, 24 participants were involved in KTHJ08 study. Later, we have shown that our experiments reveal a positive impact of the *part* feature on predicting the source and target gaze fixation duration. This feature appears in the dataset as *P01, P02, P03, P04, P05, P06, P07* and so on.

3.2 Frequecies of Unigrams and Bigrams (*Prob1, Prob2*)

Prob1 and *Prob2* are basically the measures of probabilities (using base $\log_{10}$) of unigrams and bigrams, respectively. Both of these features play significant roles in our experiments. The *Prob1* contains the values like *−4.8027, −1.4170, −4.1858* etc. whereas *Prob2* contains values like *0, −6.2460, −5.9008* etc.

3.3 Parts of Speech (*POS*)

Parts of Speech plays very crucial role in various text processing activities. In our experiments, we did not find any such exception. Different types of *POS* are considered like *NNP, VBZ, JJ, PRP* etc. in the present study.

3.4 Length

Length of a particular source as well as target word is very important while predicting the gaze duration. It is observed that the longer words are fixed with longer durations. The number of characters belong to a particular token is considered as the length of that particular token e.g., the token *"Sociology"* contains 11 character including double quotes ("), so the effective length of this particular token is *11−2 = 9*.

3.5   Tokens in Source Alignment Unit (*SAUnbr*)

It basically produces the number of alignments between the source and target units. If the source token, '*of*' is translated to '*de*' in spanish (*of* – > *de*), SAUnbr gives score as *1* (one). On the other hand, if '*sleeping_medicine*' is translated to '*tranquilizantes*' (*sleeping_medicine* – > *tranquilizantes*), the word '*sleeping*' and '*medicine*' both are aligned to '*tranquilizantes*'. So, SAUnbr gives score *2* in this case. It is observed that this feature has significant impact on predicting the source and target gaze fixation durations. The values present in the dataset are like *1, 2, 3* etc.

3.6   Typing InEfficiency (*InEff*)

It is basically the ratio of the number of produced characters and length of the final translation.

$$InEff = \#typed\ characters / length\ of\ final\ translation$$
$$= (Inseration + Deletion) / (Insertion - Deletion + 1)$$

Generally, it has shown positive impact in both of our experimental scenario. The values appear in our dataset as *0, 1, 0.19, 0.75* etc.

3.7   Cross (*Cross*)

This feature is used for considering the word alignment information while translating from source to target language. This alignment is directional, i.e. from source to target ($Cross^S$) and from target to source ($Cross^T$) and the values are like *2, 1, −1, −2, 3* etc.

3.8   Number of Mirco Units (*Munit*)

This feature is represented as *Munit* in the dataset and indicates how many micro units have contributed to the production of an alignment unit. However, in principle, there can be any number of micro units which is defined as how often a translator can revise a piece of text. It can have several information like *Dur* which indicates it's starting time and duration of the typing activity, *pause* which indicates a pause preceding the typing activity, $Paral^S$ which is basically the amount of concurrent reading activity in the source text and $Paral^T$ is same as above but it is for the target text. Moreover, it is basically characarised by typing activity, i.e. how many edit strings are there and they are represented in values like *0, 1* etc.

3.9   Word Translation Entropy (*HTra*)

Entropy (H) is related to perplexity (P) according to the equation, $P = 2^H$. Perplexity measures the number of translation choices a translator has at a particular time, for a source text s if there are $t_{1....n}$ translation choices. On the other hand, entropy measures the probability of this translation, so entropy can be mathematically defined as,

$$H(S) = \sum_{t=1}^{n} p(s \rightarrow t) * \log_2(p(s \rightarrow t)) \tag{1}$$

Where, entropy $H(S)$ is the sum over all the observed word translation alternatives multiplied by their information content. It also represents the amount of non-reduntant information provided by each of the new translation choices. Our present experiment shows that the entropy has great impact on predicting the source and target gaze fixation duration and the values are present like *0.2108, 0, 0.6747,1.0137* etc. in the dataset.

4   Experimental Setup and Result Analysis

4.1   Experiments on Source Gaze

The dataset contains a source text (*.st*) file where the total number of instances was 10640 and the values of the gaze ranges between 0 to 27832 mm. In our experiment, we aim to predict the gaze fixation duration at source side (Trt^S). So, it is considered as a dependant variable in our experiments and several other features that were extracted from the same data set have been considered as independant variables. The distributions of the gaze values are shown in the following Fig. 1. We have used R^4, a free software environment for statistical computing and graphics and Rstudio[5], an integrated development environment (IDE) for R for this purpose. The distribution of the gaze values shows that the maximum number of instances presents with zero gaze value. The values of the gaze fixation lie between the ranges of 0 to 3111 ms. Therefore, all the source instances are taken into consideration and grouped into four buckets in such a way so that each of the buckets contains equal number of instances e.g., $\underline{G^{S1}}$: (0 ms–160 ms), $\underline{G^{S2}}$: (161 ms–475 ms), $\underline{G^{S3}}$: (476 ms–980 ms) and $\underline{G^{S4}}$: (981 ms–27832 ms), respectively.

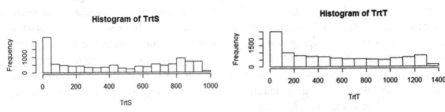

Fig. 1. Distribution of gaze values at source

Fig. 2. Distribution of gaze values at target

---

[4] https://www.r-project.org/.

[5] https://www.rstudio.com/products/rstudio/.

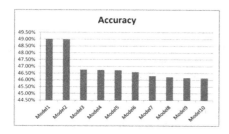

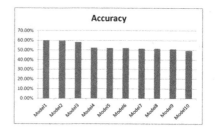

Fig. 3. 10 best accuracies in source gaze

Fig. 4. 10 best accuracy in target gaze

Table 2. Models and feature sets for source gaze and target gaze.

M^S	Source feature set (FS^S)	M^T	Target feature set (FS^T)
I	*Part*, *Prob1*, *Prob2*, *PoS*, *Length*, *SAUnbr*, *Munit*, *Edit2*, *Dur2*, *Pause2*, *ParalS2*, *Del*, *Cross*, *FFDur*	I	*Part*, *Prob1*, *Prob2*, *Length*, *PoS*, *SAUnbr*, *TAUnbr*, *Munit*, *Edit2*, *ParalT2*, *Ins*, *Del*, *Cross*, *Reg*, *FPDurT*
II	*Part*, *Prob1*, *Prob2*, *PoS*, *Length*, *SAUnbr*, *Munit*, *Edit2*, *Dur2*, *Pause2*, *ParalS2*, *Del*, *Cross*, *FFDr*, *Regr*	II	*Part*, *Prob1*, *Prob2*, *Length*, *PoS*, *SAUnbr*, *TAUnbr*, *Munit*, *Edit2*, *ParalT2*, *Ins*, *Del*, *Cross*, *Reg*, *FPDurT*, *InEff*
III	*Part*, *Prob1*, *Prob2*, *PoS*, *Length*, *SAUnbr*, *Munit*, *Edit2*, *Dur2*, *Pause2*, *ParalS2*, *Del*, *Cross*, *FFDr*, *RPDur*, *Regr*	III	*Part*, *Prob1*, *Prob2*, *Length*, *PoS*, *SAUnbr*, *TAUnbr*, *Munit*, *Edit2*, *ParalT2*, *Ins*, *Del*, *Cross*, *Reg*, *FPDurT*, *InEff*, *AltT*

Our objective is to predict the source gaze fixation duration by finding some independent features in a machine learning framework. One of the prime motivations behind these experiments is to determine the bucket class where a particular instance belongs to. So, we basically classify the instances into four classes and identify the crucial features for this classification. Weka, a machine learning tool, [10] used for this purpose, is available along with LibSvm [15]. We have extracted different features available in the dataset and combined the features to build different models, train the models using 10 fold cross validation and different classification accuracies have been obtained. As there are no such test data available, we had to undergo for the cross validation. The Fig. 3 shows the best 10 results for achieving the classification accuracies. In this case, the maximum accuracy of 49.01% has been achieved.

4.2   Experiments on Target Gaze

Similarly, in order to carry out the experiements on gaze at the target side of the language, the same experimental setup has been used. In this case, the total number of 69 files have been processed. The total number of instances was 10640 and the value of target gaze fixation duration ranges between 0 to 31939. The distribution of instances are shown in the following Fig. 2. It shows that the maximum number of instances occur with zero gaze. As all the instances are taken into account for our experiments, the instances are clustered into four clusters by keeping in mind that all groups contain equal number of instances. Under this criterion, 25% baseline accuracy is achieved by our system. The classes are G^{T1}: (0 ms–250 ms), G^{T2}: (251 ms–900 ms), G^{T3}: (901 ms–2200 ms) and G^{T4}: (2201 ms–31939 ms), respectively.

Several features have been extracted from the dataset targeting the gaze value, and combined different set of features in order to build and train different models in Weka. Several sets of experiments have been carried out based on 10-fold cross validation modes. Different classification accuracies have been achieved. The best 10 results are shown in Fig. 4 where the highest accuracy is 59.78%. The three best models and their corresponding feature sets for source and target gaze are shown in Table 2 in descending order of their results, i.e. model 1 in source and target side provides the best result, common features between two sets are marked as bold, where M^S, M^T and FS^S and FS^T indicate the models and feature sets for source and target, respectively.

The above Table shows the features like *Part, Prob1, Prob2, Length, SAUnbr* etc. which are effective to predict both source and target gaze fixation duration.

4.3 Experiments on Post Editing Dataset

The study present in TPR-DB V2.1 namely BML12 has been considered as our dataset for conducting experiments related to post editing. This study contains translating, post-editing, and editing data of six texts from English to Spanish. Particularly, the post-editing data of MT output without accessing to the source text is taken into consideration. As our aim was to predict the source gaze fixation duration here, the distributions of instances in the whole data set were plotted. The distributions are shown in Fig. 5, which clearly shows that the maximum number of instances belongs to zero gaze value. Thus, considering all the instances, we have grouped them into four classes, namely G^{PE1}, G^{PE2}, G^{PE3}, and G^{PE4}, respectively. In this section, it was observed from the distribution of dataset that the maximum number of instances belongs to zero gazes and therefore our classification is biased towards the first class. The distributions are as follows, G^{PE1} (18282): 0–2 ms, G^{PE2} (1080): 3–100 ms, G^{PE3} (2343): 101–300 ms and G^{PE4} (4711): 301–21676 ms. Different features have been extracted from the dataset and different models were built, trained on LibSvm based on 10 fold cross validation. Some of our classification results and their corresponding models are given below in Table 3.

In these experiments, only four features are taken into consideration. Table 3 shows the *Parts* feature has a great impact in this prediction. However, *Prob1* produces little improvement, but *Prob2* and *PoS* do not show any improvement in the result. So, in this case, the combination of *Part and Prob1* achieves the highest accuracy of 71.70%.

Table 3. Experimental results and feature set.

Feature set for post editing dataset (FS^{PE})	Accuracy
Information gain: Parts - 0.224, *Prob1* - 0.1564, *Prob2* - 0.094, *PoS* - 0.0037.	
Parts	70.08%
Part, Prob1, Prob2, PoS	70.72%
Part, Prob1	**71.70%**
Part, Prob1, Prob2	**71.36%**
Part, Prob1, PoS	71.35%

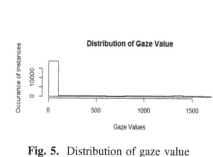

Fig. 5. Distribution of gaze value

The calculation of Pearson product-moment correlation coefficient for each of the four features has also been performed by an ablation study which is depicted in Table 4.

Table 4. Feature and it's corresponding Pearson correlation coefficient

Feature: Pearson correlation coefficient
Part: 0.3585, ***prob1:*** *0.0731,* ***prob2:*** *0.0414,* *Pos: 0.1288*

In order to improve the performance of the system, in this case, we calculated the *information gain*[6] for each of the features. Information gain is basically related to entropy which is nothing but the measure of information contained to that particular variable. The higher the entropy, the more is the information content.

5   Information Gain Based Pruning

Information Gain tells us how important a given attribute or feature is within a feature vector. Thus, we generally use it to decide the ordering of the attributes in a decision tree.

5.1   On Translation Data

Information gain of each feature on the translation data has been calculated and the values of some of the features have been shown in the Table 5.

Table 5. Feature and it's corresponding information gain

Feature: information gain
Part: −0.333565070484262, ***prob1:*** *−0.317373638365639,* ***prob2:*** *−0.371524896696689,* *Pos:* −0.282699940024125, ***SAUnbr:*** *−0.38341195641251,* ***TAUnbr:*** *−0.37946753724893,* *Cross:* −0.399714167594243

---

[6] http://homes.cs.washington.edu/ ∼ shapiro/EE596/notes/InfoGain.pdf.

5.2   On Post Edited Data

Translator's Identity: In the dataset, 32 translators are mentioned namely *p01, p02, p03....p0n* and so on. We have calculated the information gain for each of the features while predicting the gaze duration. To calculate the information gain for this particular feature, we assumed that 32 distinct children are emerged and each child has certain number of instances belonging to any of the four classes (G^{PE1}, G^{PE2}, G^{PE3}, and G^{PE4}). We have calculated the entropy for each child i.e. for *p01, p02....p0n* and so on. To calculate entropy say for *p01* following equation is employed.

$$Entropy_{p01} = - \frac{nG^{PE1}}{np_{01}} * \log_2 \frac{nG^{PE1}}{np_{01}} - \frac{nG^{PE2}}{np_{01}} * \log_2 \frac{nG^{PE2}}{np_{01}} - \frac{nG^{PE3}}{np_{01}} * \log_2 \frac{nG^{PE3}}{np_{01}} - \frac{nG^{PE4}}{np_{01}} * \log_2 \frac{nG^{PE4}}{np_{01}}$$

Where, nG^{PEi} = number of instances in Gi class for $i = 1$ to 4, np_{01} = total number of instnaces in child *p01*. We have calculated the entropy for p_{oi} where i = 0, 1,...32. We calculate the weighted entropy, i.e. by multiplying each entropy with total number of instances for that group normalized by the total number of instances present in the dataset. Thus, we take the sum of each such weighted entropy of all the children, and finally we achieved the information gain for that particular feature i.e. for translator's identity by substracting the sum of the weighted entropy from 1.

Unigram Probability: The distribution of unigram probabilities is shown in Fig. 6. The histogram shows that the values are either less than zero or less than −1.3096 or greater than −50. The diagram also shows that the maximum number of instances lies within the range of −1 to −5. We have analysed the data in deeper thought, i.e. we have calculated the number of instances belong to a particular range, and also calculated the number of instances belonging to each class (i.e. in G^{PE1}, G^{PE2}, G^{PE3}, and in G^{PE4}). It is observed from the dataset that the number of instances in G1 is much more than others, as beacuse the number of instance with zero value is very huge in the original dataset. We have further classified the the total number of instaces into G^{PE1} and other than G^{PE1} in a certain range of values as the number of instances presents in G^{PE1} is much more compared to remaining three classes and is depicted in Table 7.

Table 6. Analysis of data value

# Total instances in *Prob1 (Prob2)*					
Value	Corpus	G1	G2	G3	G4
−1.3096 (−2.2331)	1529 (190)	683 (139)	64 (5)	169 (14)	613 (32)
−2.055 (−3.0006)	219 (32)	47 (23)	9 (3)	46 (1)	117 (5)
−3.0056 (−5.0094)	91 (30)	0 (25)	0 (1)	0 (1)	0 (3)
−4.2514 (−7.0497)	31 (246)	13 (174)	5 (12)	8 (22)	5 (38)
−50 (−50)	91 (4087)	60 (2706)	2 (151)	11 (369)	18 (861)

Table 7. Value range wise classes

Ranges	# of instances prob1 (prob2)	# instances in G1 prob1 (prob2)	# instances other than G1 prob1 (prob2)
$-1 <$ val < -3	11953 (970)	6521 (703)	5432 (267)
$-3 <$ val < -5	11551 (6968)	9594 (5088)	1957 (1880)
$-5 <$ val < -7	2715 (11221)	2032 (7712)	683 (3509)
$-7 <$ val < -50	197 (7257)	135 (4779)	62 (2478)

Bi-Gram Probability: Similarly, the distributation of the values for this particular feature is shown in Fig. 7. The diagram shows that the maximum number of instances belongs to values of 0 to -10. We make deep analysis on the values of this feature and identified 468 unique values. We also calculated the number of instances belonging to a particular unique value, and thus, we have further calculated the number of instances with G^{PE1}, G^{PE2}, G^{PE3} and G^{PE4} classes. The Table 6 shows such analysis with some selected values out of total 468 unique values.

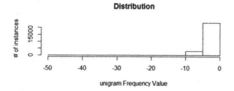

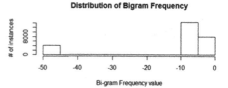

Fig. 6. Distribution of unigram probability

Fig. 7. Distribution of bigram probability

Further, we identified the number of instances belonging to $\mathbf{G^{PE1}}$ and the number of instances belonging to other than $\mathbf{G^{PE1}}$ for the same reason as mention in case of unigram probability. Thus, we made grouping on the whole set of values like shown in the Table 7. We have calculated the total number of instances, number of instances with group G1 and the number of instances in other group.

Parts of Speech (PoS): The dataset contains 28 distinct parts of speech (PoS) type. We have calculated the information gain for this particular feature. It can be thought that the total number of instances is parent and there are 28 number of children that are emerged from this parent. We have calculated entropy for each of the children. So, the following equation has been governed to calculate the entropy of a particular child.

$$Entropy_{Child} = -\sum \frac{nG^{PEi}}{nc} * \log_2 \frac{nG^{PEi}}{nc}$$

Where, nG^{PEi} = number of occurance of class and nc = number of occurance of child class. The entropy of each child has been identified and therefore, we calculated the weighted entropy of that child by multiplying it with the total number of instances belong to that child normalized by the total number of instances present in whole corpus. Once such calculation is done, the summation of each child's weighted entropy has been taken into consideration. Finally, the information gain is obtained by substracting the weighted sum from 1.

6 Conclusion and Future Work

Several conclusions can be drawn from the different sets of experiments carried out in the present task. We have considered two different dataset, one with translation data and other with post edited data employed in machine learning framework. In the first one, experiments are subdivided further into two part, i.e. one with source gaze other with target gaze. Different classification accuracies have been achieved in both the cases. The features which play important role are translator's identity, unigram and bigram probability, parts of speech, cross and many more. In the second one, four sets of experiments have been carried out and their models and classification accuracies are shown. In this case, all the features except *Parts of Speech (PoS)* and *Bi-gram Probability (Prob2)* have shown their contributions in predicting source gaze fixation duration. At the end, information gain based pruning of each feature extracted from two datasets has been carried out. The effectiveness of the feature has been reevaluated using information gain also.

Acknowledgements. The research work has received funding from the project "Development of Tree Bank in Indian Languages (TBIL)" funded by The Department of Electronics and Information Technology (DeitY), Ministry of Communication and Information Technology, Government of India.

References

1. Carl, M.: Translog-II: a program for recording user activity data for empirical reading and writing research. In: Proceedings of the Eight International Conference on Language Resources and Evaluation, European Language Resources Association (ELRA) (2012)
2. Jakobsen, A.L., Schou, L.: Translog documentation. In: Probing the Process in Translation: Methods and Results. Copenhagen Studies in Language Series, vol. 24, pp. 1–36 (1999)
3. Carl, M., Schaeffer, M.J.: The CRITT translation process research database V1.4. In: The Bridge: Research Platform. Department of International Business Communication (IBC). Copenhagen Business School (2014)
4. Carl, M.: The CRITT TPR-DB 1.0: a database for empirical human translation process research. In: AMTA 2012 Workshop on Post-Editing Technology and Practice (WPTP-2012) (2012)
5. Prasov, Z., Chai, J.Y.: What's in a gaze?: the role of eye-gaze in reference resolution in multimodal conversational interfaces. In: Proceedings of the 13th International Conference on Intelligent User Interfaces, pp. 20–29 (2008)
6. Meyer, A.S., et al.: Viewing and naming objects: eye movements during noun phrase production. Cognition **66**, B25–B33 (1998)

7. Rayner, E.D., et al.: Toward a model of eye movement control in reading. Psychol. Rev. **105**, 125–157 (1998)
8. Rayner, K., McConkie, G.W.: What guides a reader's eye movements? Vis. Res. **16**(8), 829–837 (1976)
9. Jakobsen, A.L., Jensen, K.T.H.: Looking at Eyes Eye-Tracking Studies of Reading and Translation Processing, vol. 36, pp. 103–124. Samfundslitteratur, Copenhagen (2008)
10. Witten, I.H., Frank, E.: Data Mining: Practical Machine Learning Tools and Techniques. Morgan Kaufmann, Burlington (2005)
11. Rayner, K., McConkie, G.W.: What guides a reader's eye movements? Vis. Res. **16**, 829–837 (1976)
12. Inhoff, A.W., Rayner, K.: Parafoveal word processing during eye fixations in reading: effects of word frequency. Percept. Psychophys. **40**(6), 431–439 (1986)
13. Carl, M., et al.: D1.3: Final report on user interface studies, cognitive and user modelling (2014)
14. Saikh, T., Bangalore, S., Carl, M., Bandyopadhyay, S.: Predicting source gaze fixation duration: a machine learning approach. In: International Conference on Cognitive Computing and Information Processing. IEEE (2015)
15. Chang, C.-C., Lin, C.-J.: LIBSVM: a library for support vector machines. ACM Trans. Intell. Syst. Technol. (TIST) **2**(3), 27 (2011)
16. Sedgwick, P., et al.: Pearson's correlation coefficient. BMJ **345**, e4483 (2012)
17. Solorio, T., Liu, Y.: Part-of-speech tagging for English-Spanish code-switched text. In: Empirical Methods on Natural Language Processing, EMNLP-2008, Honolulu, Hawaii, pp. 1051–1060, October 2008
18. Li, Y., et al.: Learning to Predict Gaze in Egocentric Video. In: Proceedings of the IEEE International Conference on Computer Vision (ICCV 2013), pp. 3216–3223 (2013)
19. Prasov, Z., Chai, J.Y.: What's in a gaze?: the role of eyegaze in reference resolution in multimodal conversational interfaces. In: Proceedings of the 13th International Conference on Intelligent User Interfaces, pp. 20–29 (2008)

Unsupervised Domain Ontology Learning from Text

Sree Harissh Venu[✉], Vignesh Mohan, Kodaikkaavirinaadan Urkalan, and Geetha T.V.

Department of Computer Science and Engineering,
College of Engineering, Guindy, India
vharissh14@gmail.com

Abstract. Construction of Ontology is indispensable with rapid increase in textual information. Much research in learning Ontology are supervised and require manually annotated resources. Also, quality of Ontology is dependent on quality of corpus which may not be readily available. To tackle these problems, we present an iterative focused web crawler for building corpus and an unsupervised framework for construction of Domain Ontology. The proposed framework consists of five phases, Corpus Collection using Iterative Focused crawling with novel weighting measure, Term Extraction using HITS algorithm, Taxonomic Relation Extraction using Hearst and Morpho-Syntactic Patterns, Non Taxonomic relation extraction using association rule mining and Domain Ontology Building. Evaluation results show that proposed crawler outweighs traditional crawling techniques, domain terms showed higher precision when compared to statistical techniques and learnt ontology has rich knowledge representation.

Keywords: Iterative Focused Crawling · Domain Ontology · Domain terms extraction · Taxonomy · Non Taxonomy

1 Introduction

Ontology in computer science can be viewed as formal representation of knowledge pertaining to particular domain [18]. In simpler terms ontology provides concepts and relationship among concepts in a domain. Machines perceive contents of documents (blogs, articles, web pages, forums, scientific research papers, e-books, etc.) as sequence of character. Much of the semantic information are already encoded in some form or other in these documents. There is an increasing demand to convert these unstructured information into structured information. Ontology plays a key role in representing the knowledge hidden in these texts and make it human and computer understandable.

Construction of Domain Ontology provide various semantic solution including: (1) Knowledge Management (2) Knowledge Sharing (3) Knowledge Organization (4) Knowledge Enrichment.

© Springer International Publishing AG 2017
R. Prasath and A. Gelbukh (Eds.): MIKE 2016, LNAI 10089, pp. 132–143, 2017.
DOI: 10.1007/978-3-319-58130-9_13

It can be effectively used in semantic computing applications ranging from Expert Systems [16], Search Engines [22], Question and Answering System [7], etc. to solve day to day problems. For example, if the search engine is aware that "prokaryote" is a type of organism, better search results can be obtained and recall of the system will be improved subsequently.

Ontology is generally built under the supervision of domain experts and are time intensive process. Corpus required for building Ontology are not always readily available. Therefore, it is important to build corpus from web through crawling. Very few work is available that have incorporated crawling as a phase for collecting corpus in building Ontology. Since general crawling does not always provide domain related pages, lot of irrelevant pages are downloaded and filtering is required. Terms extracted using statistical measure or linguistic patterns are prone to noise and require additional level of filtering using machine learning techniques. Also, most systems rely on manually annotated resources for obtaining terms and also for relation discovery. These resources however mostly contain domain generic concepts and lack domain specific concepts and relations [18]. Ontology extracted using lexico-syntactic patterns are limited to certain patterns and require enrichment.

In this work we propose a framework for crawling websites relevant to the domain of interest and also build Domain Ontology without use of any annotated resource in an unsupervised manner. The crawling framework uses a novel weighting measure to rank the domain terms. The proposed framework consists of five phases Corpus Collection, Term Extraction, Taxonomic Relation Extraction, Non-taxonomic relation extraction and Domain Ontology building. Corpus is crawled using iterative focused web crawler which downloads the content which are pertinent to the domain by selectively rejecting URL's based on link, anchor text and link context. Terms are extracted by feeding graph based algorithm HITS with Shallow Semantic Relations and proposed use of adjective modifiers to obtain fine grained domain terms. Hearst pattern and Morpho-Syntactic patterns are extracted to build taxonomies. Non-taxonomic relation extraction is obtained through Association Rule Mining on Triples.

The organization of the paper is as follows: Sect. 2 describes Related Work, Sect. 3 describes the System Design, Sect. 4 describes the Results and Evaluation, Sect. 5 describes the Conclusion and Future Work.

2 Related Work

In this section, we discuss the literature survey in Corpus Collection, Term Extraction, Taxonomic Relation Extraction and Non Taxonomic Relation Extraction.

2.1 Domain Corpus Collection

Domain Corpus is a coherent collection of domain text. It requires the usage of iterative focused or topical web crawler to fetch the pages that are pertinent to the domain of interest. In the work proposed by [6], a heuristic based approach is used to locate anchor text by using DOM tree instead of using the entire HTML Page. A statistical based term weighing measure based on TF-IDF called TFIPNDF (Term Frequency Inverse Positive Negative Document Frequency) was proposed for weighing anchor text and link context. The pages are classified as relevant or not relevant on the basis of trained classifier and is entirely supervised. The work however lacks iterative learning of terms to classify pages [15].

2.2 Domain Term Extraction

Domain Terms are the elementary components used to represent concepts of a domain. Example of domain terms pertaining to agricultural domain are "farming", "crops", "plants", "fertilizers", etc. Term Extraction is generally performed from collection of domain documents using any of the following methods: Statistical Measure, Linguistic Measure, Machine Learning and Graph-based Measure.

Statistical Measure. Most common Statistical Measure make use of TF (Term Frequency) and IDF (Inverse Document Frequency). Meijer et al. [9], proposed four measures namely Domain Pertinence, Lexical Cohesion, Domain Consensus and Structural Relevance to compute the importance of terms in a domain. Drymonas et al. [3], used C/NC values to calculate the relevance of multiword terms in corpus. These measures however fail to consider the context of terms and fails to capture the importance of infrequent domain terms.

Linguistic Measure. Linguistic Measures traditionally acquire terms by using syntactic patterns such as Noun-Noun, Adjective-Noun, etc. For example, the POS tagging of the sentence "Western Rajasthan and northern Gujarat are included in this region" tags "Western" as an adjective and "Rajasthan" as Noun. Lexico-Syntactic patterns makes use of predefined patterns such as "including", "like", "such as", etc., to extract terms. It is however tedious and time consuming to pre-define patterns.

Machine Learning. Machine Learning is either supervised or unsupervised. Supervised learning require the algorithm to be trained before usage and target variable is known. Some famous and commonly used supervised algorithms include Naive Bayes, Support Vector Machines and Decision Tree. In unsupervised learning training is not required and hidden patterns are found using unlabeled data. Uzun [21] work considers training features are independent and therefore used TF-IDF, distance of the word to the beginning of the paragraph, word position with respect to whole text and sentence and probability features from Naive Bayes Classifier to classify whether a term is relevant. The drawback of using machine learning is that training incurs overhead and data may not be available in abundance for training.

Graph Based Measure. Graph Based Measure is used to model the importance of a term and the relationship between the terms in an effective way. Survey on Graph Methods by Beliga et al. [1], suggest that graphs can be used to represent co-occurrence relations, semantic relations, syntactic relations and other relations (intersecting words from sentence, paragraph, etc.). Work by Ventura et al. [8] used novel graph based ranking method called "Terminology Ranking Based on Graph Information" to rank the terms and dice coefficient was used to measure the co-occurrence between two terms. Mukherjee et al. [10] used HITS index with hubs as Shallow Semantic Relations and authorities as nouns. Terms are filtered based on hubs and authority scores.

2.3   Taxonomic Relation Extraction

Taxonomy construction involves building a concept hierarchy in which broader-narrower relations are stored and can be visualized as a hierarchy of concepts. For example "rice", "wheat", "maize" come under "crop". They are commonly built using predefined patterns such as the work by Hearst [4] and Ochoa [12] et al. Meijer et al. [9] proposed construction of taxonomy using subsumption method. This method calculates co-occurrence relations between different concepts. Knijff et al. [2], compared two methods subsumption method and hierarchical agglomerative clustering to construct taxonomy. They concluded that subsumption method is suitable for shallow taxonomies and hierarchical agglomerative clustering is suitable for building deep taxonomies.

2.4   Non Taxonomic Relation Extraction

Non Taxonomic Relations best describe the non-hierarchical attributes of concept. For example, in the non taxonomic relation "predators eat plants", eat is a feature of predator. Nabila et al., [11] proposed an automatic construction of non-taxonomic relation extraction by finding the non-taxonomic relations between the concepts in the same sentence and non-taxonomic relations between concepts in different sentences. Serra and Girardri [14] proposed a semi-automatic construction of non-taxonomic relations from text corpus. Association between two concepts are found by calculating the support and the confidence scores between the two concepts.

To build a Domain Ontology from Text, the existing methods for Domain Term Extraction deprive from identification of low frequent terms, identification of all syntactic-patterns and require annotated re-sources for machine learning approaches. Graph based methods for identification can be used to solve the above problems as they can represent the meaning as well as composition of text. They also do not require manually annotated data unlike machine learning approaches. General Non-Taxonomic Relation Extraction methods are based on extraction of predicates between two concepts and as all predicates are not domain specific the use of Data Mining Techniques can be helpful in identifying the Domain Relations effectively.

3 System Design

In this section we discuss the design of our system. Figure 1 shows the overall architecture diagram of the proposed framework. The system consists of five major phases: (1) Domain Corpus Collection (2) Domain Term Extraction (3) Taxonomic Relation Extraction (4) Non Taxonomic Relation Extraction and (5) Domain Ontology Building.

3.1 Domain Corpus Collection

Corpus required for construction of Ontology may not be readily available for every domain. Since the quality of the corpus plays a vital role in deciding the quality of Ontology, Iterative Focused Crawling is performed to download web pages relevant to the domain. List of Seed URLs are given as input to the Iterative Focused Crawler. The web pages whose URL, anchor text or link context satisfy the relevance score are added to the URL queue. The depth of the pages to be crawled is specified. The output of the focused crawler is used as corpus for construction of Ontology. Crawling is terminated when the relevance of URL to the context vector decreases drastically. The architecture of crawler is depicted in Fig. 2.

Nouns are considered as candidate terms for finding keywords in the domain. Therefore, the nouns are extracted from the corpus using the Stanford parts-of-speech tagger. The context vector of a noun is computed by using proposed weighted co-occurrence score. Weighted co-occurrence ($WCO(w_i, w_j)$) of two words w_i and w_j is given by:

$$WCO(w_i, w_j) = CO(w_i, w_j) X idf(w_i) X idf(w_j) \tag{1}$$

In Eq. 1, $idf(w_i)$ and $idf(w_j)$ are the inverse document frequency of words w_i and w_j. $CO(w_i, w_j)$ is the co-occurrence frequency of the two words w_i and w_j. The proposed equation considers the inverse document frequencies of the terms in order to consider the importance of terms which occur rarely and may of importance to the domain. Unit Normalization of the context vector is performed to have a specific range of score between 0 and 1. The normalized context vector of each term is summed along the column and sorted in descending order. The top ranked terms are extracted as concepts based on percentage.

Relevance of the web pages are calculated by computing the average of the Cosine Similarity Score of the test domain vectors and each of the domain vectors of the training document. The relevance of the URL is checked without scanning the pages. It is done by computing relevance of HREF, Anchor Text and/or Link Context. Appropriate threshold are set for HREF, Anchor Text and Link Context. If HREF is not relevant (i.e. Relevance Score), Anchor Text will be checked for relevance. If Anchor Text is not relevant, finally, Link Context will be checked.

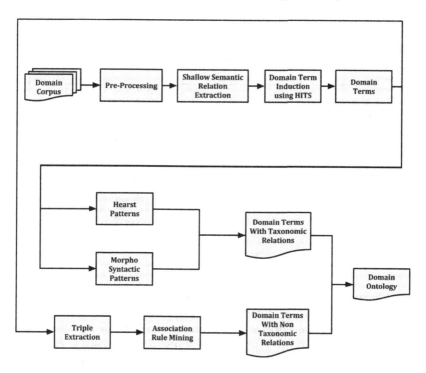

Fig. 1. Architecture of proposed framework: unsupervised domain ontology construction from text

3.2 Domain Term Extraction

Domain corpus, which contains a rich collection of text documents is pre-processed to identify the domain terms. Numbers, special characters, etc. which do not play a significant role in ontology construction are removed.

Shallow Semantic Relation Extraction. Domain text documents are tokenized into sentences. These sentences are parsed using Stanford Dependency Parser to identify the Shallow Semantic Relations between the words. Shallow Semantic Relations represent the syntactic contextual relations within the sentences. In addition to the Shallow Semantic Relations extracted in [10] we have also extracted and used adjective modifiers obtained through Dependency Parsing. Since, significant amount of domain terms are composed as adjective modifier, it is important to consider these dependencies. For example, in the sentence "Biological research into soil and soil organisms has proven beneficial to organic farming.", "organic farming" and "biological research" are tagged as adjective modifiers.

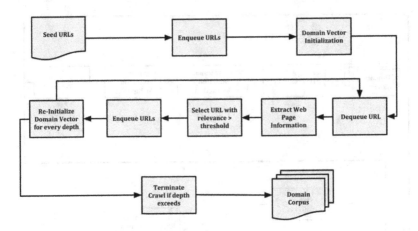

Fig. 2. Flow diagram of iterative focused crawler

Domain Term Induction Using HITS. HITS algorithm [5,10] is applied to identify the most important domain terms. It is composed of two major components – Hubs and Authorities. Hubs are represented by Shallow Semantic Relations and authorities are represented by nouns. Hub score is calculated as the sum of authority scores and authority score is calculated as the sum of hub scores. Hub and Authority score are calculated recursively until hub and authority score converges. The Shallow Semantic Relation which has high hub score are selected as multi-grams and nouns which has high authority score are selected as unigrams. These unigrams and multi-grams constitute the domain terms.

3.3   Taxonomic Relation Extraction

Taxonomic Relations represent hypernym-hyponym relation. A hypernym represents the specific semantic field of a hyponym and a hyponym represents the generic semantic field of the hyponym. The three steps involved in building a taxonomy involves (i) Hearst Pattern Extraction and (ii) Morpho-syntactic Pattern Extraction.

Hearst Pattern Extraction. Hearst Patterns [4] are commonly used to extract taxonomic relations from text. In our work we leverage rule based technique presented in the above mentioned paper to induce taxonomy. Sentences containing the domain terms are selected for identification of Hearst Patterns. Sentences are tagged using parts-of-speech tagger to find taxonomic relations.

Morpho Syntactic Pattern Extraction. In our work we have also extracted Morpho Syntactic Patterns [12] to extract additional Hypernym-Hyponym relations. There are two rules followed to extract morpho-syntactic patterns.

Rule 1: If the term t_1 contains a suffix string t_0, then the term t_0 is the hypernym of the term t_1, provided the term t_0 or t_1 is a domain term. For example, "polysaccharide" is considered as the hypernym of the term "homopolysaccharide".

Rule 2: If the term t_0 is the head term of the term t_1, then t_0 is considered as the hypernym of the term t_1, provided term t_0 or t_1 is the domain term. Example: "sweet corn" is the hyponym of the word "corn".

3.4   Non-Taxonomic Relation Extraction

Non-Taxonomic Relations represent the properties of the object. It has no class-subclass relationship.

Triplet Extraction. A sentence is composed of three components - subject, predicate and object. A triplet in a sentence is defined as the relation between the subject and the object, with the relation being the predicate. Parsed documents using Stanford Parser are input to the triplet extraction process. Subject, predicate and object from the sentences is extracted using Russu's Triple Algorithm [13].

Association Rule Mining. Association Rule Mining [17] is performed to find the non-taxonomic relations between the domain terms. Apriori Algorithm is used for frequent itemset generation and association rule mining. Frequent itemset whose support crosses a suitable threshold are selected for mining association rules. Association rules are filtered from frequent itemsets and association rules which satisfy a suitable confidence score are selected.

3.5   Domain Ontology Building

The concepts with the taxonomic and non-taxonomic relations are represented in a Resource Description Framework format. The concepts consists of a concept id, a broader relation, a narrower relation and a non-taxonomic relation associated with it. The broader/narrower relation are represented by class/subclass relations. Non-taxonomic relations consists of a property, domain and range. The domain of a property represent the subject whose predicate is that property. The range of a property represent the object whose predicate is that property. Example: "rice" is a concept with concept id "12143", narrower relations "long-grain rice", broader relation "crops", "medium-grain rice", "short-grain rice", property "grows in", domain "rice", range "South India".

4   Results and Evaluation

4.1   Domain Corpus Collection

Domain Corpus Collection consists of implementing an iterative focused web crawler that crawls pages relevant to the domain. 22 seed URLS pertaining to

Table 1. Number of links crawled through HREF, anchor text and link context

Mode	Count
HREF	606
Anchor text	2256
Link context	17842
Total	20632

Table 2. Number of documents in different similarity range

Similarity	Count
0.6–0.7	787
0.5–0.6	11624
0.4–0.5	5782
0.3–0.4	2156
0.2–0.3	251
0.1–0.2	60
0.0–0.1	22
Total	20632

agriculture domain were given as input to the focused crawler. 20,632 documents were obtained at the end of crawling a depth of 3. 22 relevant links were crawled in depth 0, 134 relevant links were crawled in depth 1, 816 relevant links were crawled in depth 2 and 19732 relevant links were crawled in depth 3.

Table 1 shows the Number of Links crawled through HREF, Anchor Text and Link Context. It is observed that most of the links were found to be relevant through HREF and Link Context. HREF usually contain the text present in the Anchor Text. So, if the relevance fails through HREF there is a high probability of checking the Link Context. Table 2 shows the Number of documents in different similarity range compared to SeedURL pages. It can be seen that most of the pages similarity were in the range of 0.5 to 0.6. It was also observed that the median of relevance score follows a decreasing trend and the number of irrelevant links crawled increased after a depth of 3. In our work, Convergence Score [20] was used to evaluate the Iterative Focused Crawler. It is defined as the number of concepts present in the final crawl to the number of concepts present in initial seed page set and has score range between 0 and 1. The convergence score was evaluated to be 0.2 and 0.43 for baseline crawling and proposed focused crawling respectively. It can be inferred that the proposed crawling mechanism was twice more effective than traditional baseline crawling approaches.

4.2 Domain Term Extraction

The precision scores of Graph Based Domain Term Extraction using HITS algorithm used in our work is evaluated against statistical measures such as Linguistic Patterns, Inverse Document Frequency, C-value (LIDF score) and Graph Based Algorithm Terminology Ranking Based on Graph Information - TeRGraph proposed by [8] and sum of statistical scores obtained from Domain Pertinence (DP), Domain Consensus (DC), Lexical Cohesion (LC) and Structural Relevance (SR) proposed in [9] is shown in Table 3. GENIA corpus used in [8] was used for evaluation purpose. The measures shows that graph based HITS

Table 3. Precision scores of term extraction using HITS, LIDF, TeRGraph and DP+DC+LC+SR

Total terms	Term extraction using HITS	LIDF	TeRGraph	DP+DC+LC+SR
1000	0.772	0.697	0.769	0.751
2000	0.749	0.662	0.694	0.687
3000	0.733	0.627	0.644	0.657
4000	0.703	0.608	0.593	0.612
5000	0.676	0.575	0.562	0.583

algorithm shows better precision compared to statistical measures and Graph Based algorithm TeRGraph.

4.3 Domain Ontology

Hearst Patterns and Morpho-Syntactic patterns were used to induce Taxonomy. Total of 6539 Hearst Patterns and 2149 Morpho-Syntactic patterns were extracted to construct the Taxonomy. 5216 triples were extracted and 357 Non Taxonomic Relations were identified using Association Rule Mining. In our work, Domain Ontology was evaluated using Metic Based Evaluation techniques Inheritance Richness and Class Richness [19].

Class Richness. This metric is related to how instances are distributed across classes. The number of classes that have instances in the KB is compared with the total number of classes, giving a general idea of how well the KB utilizes the knowledge modeled by the schema classes. Low Class Richness implies KB does not have data that exemplifies all the class knowledge that exists in the schema. High CR would indicate that the data in the KB represents most of the knowledge in the schema. Table 4 shows the Class Richness score for Taxonomy and Non Taxonomy learning methods.

Inheritance Richness. Inheritance Richness measure describes the distribution of information across different levels of the ontology's inheritance tree or the fan-out of parent classes. This is a good indication of how well knowledge is grouped into different categories and subcategories in the ontology. This measure can distinguish a horizontal ontology (where classes have a large number of direct subclasses) from a vertical ontology (where classes have a small number of direct subclasses). Table 4 shows the Class Richness score for Taxonomy and Non Taxonomy learning methods. From the results of the evaluation metrics (class richness and inheritance richness), it is evident that the constructed ontology has a good density depicting that the concepts extracted represents a wider knowledge in the domain.

Table 4. Inheritance and class richness scores

Method	Inheritance richness	Class richness
Hearst	4.004	0.367
Morpho-syntactic	2.671	0.068
Hearst + Morpho-syntactic	3.967	0.41
Non-taxonomic relation	1.81	0.21

5 Conclusion and Future Work

In our work, we have developed an iterative focused crawler for collection of domain corpora, with each element in the co-occurrence matrix weighted as product of co-occurrence frequency and IDF of row and column. Domain terms were extracted without any manual annotated resource unsupervised using HITS algorithm with Hubs as Shallow Semantic Relation and Authority as Nouns. The ranked terms were removed of noise using Domain Pertinence. In this work, taxonomy was induced using Hearst Patterns and Morpho-Syntactic Patterns. The Ontology was built automatically without supervision from scratch. In the future, we intend to exploit deep learning methods for building Domain Ontology to make it meaningful and useful.

References

1. Beliga, S., Meštrović, A., Martinčić-Ipšić, S.: An overview of graph-based keyword extraction methods and approaches. J. Inf. Organ. Sci. **39**(1), 1–20 (2015)
2. De Knijff, J., Frasincar, F., Hogenboom, F.: Domain taxonomy learning from text: the subsumption method versus hierarchical clustering. Data Knowl. Eng. **83**, 54–69 (2013)
3. Drymonas, E., Zervanou, K., Petrakis, E.G.M.: Unsupervised ontology acquisition from plain texts: the *OntoGain* system. In: Hopfe, C.J., Rezgui, Y., Métais, E., Preece, A., Li, H. (eds.) NLDB 2010. LNCS, vol. 6177, pp. 277–287. Springer, Heidelberg (2010). doi:10.1007/978-3-642-13881-2_29
4. Hearst, M.A.: Automatic acquisition of hyponyms from large text corpora. In: Proceedings of the 14th Conference on Computational Linguistics, vol. 2, pp. 539–545. Association for Computational Linguistics (1992)
5. Kleinberg, J.M.: Authoritative sources in a hyperlinked environment. J. ACM (JACM) **46**(5), 604–632 (1999)
6. Liu, L., Peng, T., Zuo, W.: Topical web crawling for domain-specific resource discovery enhanced by selectively using link-context. Proc. Int. Arab J. Inf. Technol. **12**(2), 196–204 (2015)
7. Lopez, V., Pasin, M., Motta, E.: AquaLog: an ontology-portable question answering system for the semantic web. In: Gómez-Pérez, A., Euzenat, J. (eds.) ESWC 2005. LNCS, vol. 3532, pp. 546–562. Springer, Heidelberg (2005). doi:10.1007/11431053_37

8. Lossio-Ventura, J.A., Jonquet, C., Roche, M., Teisseire, M.: Yet another ranking function for automatic multiword term extraction. In: Przepiórkowski, A., Ogrod-niczuk, M. (eds.) NLP 2014. LNCS (LNAI), vol. 8686, pp. 52–64. Springer, Cham (2014). doi:10.1007/978-3-319-10888-9_6

9. Meijer, K., Frasincar, F., Hogenboom, F.: A semantic approach for extracting domain taxonomies from text. Decis. Support Syst. **62**, 78–93 (2014)

10. Mukherjee, S., Ajmera, J., Joshi, S.: Domain cartridge: unsupervised framework for shallow domain ontology construction from corpus. In: Proceedings of the 23rd ACM International Conference on Conference on Information and Knowledge Management, pp. 929–938. ACM (2014)

11. Nabila, N., Mamat, A., Azmi-Murad, M., Mustapha, N.: Enriching non-taxonomic relations extracted from domain texts. In: 2011 International Conference on Semantic Technology and Information Retrieval, pp. 99–105. IEEE (2011)

12. Ochoa, J.L., Almela, Á., Hernández-Alcaraz, M.L., Valencia-García, R.: Learning morphosyntactic patterns for multiword term extraction. Sci. Res. Essays **6**(26), 5563–5578 (2011)

13. Rusu, D., Dali, L., Fortuna, B., Grobelnik, M., Mladenic, D.: Triplet extraction from sentences. In: Proceedings of the 10th International Multiconference Information Society-IS, pp. 8–12 (2007)

14. Serra, I., Girardi, R.: A process for extracting non-taxonomic relationships of ontologies from text (2011)

15. Gangly, B., Sheikh, R.: A review of focused web crawling strategies. Int. J. Adv. Comput. Res. **2**(4) (2012)

16. Shue, L.Y., Chen, C.W., Shiue, W.: The development of an ontology-based expert system for corporate financial rating. Expert Syst. Appl. **36**(2), 2130–2142 (2009)

17. Srikant, R., Agrawal, R.: Mining generalized association rules. IBM Research Division (1995)

18. Sure, Y., Staab, S., Studer, R.: Ontology engineering methodology. In: Staab, R., Studer, R. (eds.) Handbook on Ontologies. International Handbooks on Information Systems, pp. 135–152. Springer, Heidelberg (2009). doi:10.1007/978-3-540-92673-3_6

19. Tartir, S., Arpinar, I.B., Moore, M., Sheth, A.P., Aleman-Meza, B.: Ontoqa: metric-based ontology quality analysis (2005)

20. Thenmalar, S., Geetha, T.: The modified concept based focused crawling using ontology. J. Web Eng. **13**(5–6), 525–538 (2014)

21. Uzun, Y.: Keyword extraction using naïve bayes. Bilkent University, Department of Computer Science, Turkey (2005). www.cs.bilkent.edu.tr/~guvenir/courses/CS550/Workshop/Yasin_Uzun.pdf

22. Zhang, Y., Vasconcelos, W., Sleeman, D.: Ontosearch: an ontology search engine. In: Bramer, M., Coenen, F., Allen, T. (eds.) Research and Development in Intelligent Systems XXI. Springer, London (2005). doi:10.1007/1-84628-102-4_5

Regression Based Approaches for Detecting and Measuring Textual Similarity

Sandip Sarkar[1](✉), Partha Pakray[2](✉), Dipankar Das[1](✉), and Alexander Gelbukh[3](✉)

[1] Jadavpur University, Kolkata, India
sandipsarkar.ju@gmail.com, dipankar.dipnil2005@gmail.com
[2] National Institute of Technology, Aizawl, Mizoram, India
parthapakray@gmail.com
[3] Instituto Politécnico Nacional, Mexico City, Mexico
gelbukh@gelbukh.com

Abstract. Finding Semantic similarity is an important component in various fields such as information retrieval, question-answering system, machine translation and text summarization. This paper describes two different approaches to find semantic similarity on SemEval 2016 dataset. First method is based on lexical analysis whereas second method is based on distributed semantic approach. Both approaches are trained using feed-forward neural network and layer-recurrent network to predict the similarity score.

1 Introduction

Semantic textual similarity (STS) measures the similarity between the two text sequences. Since 2013 SemEval workshop attracts researchers from many research groups. Like previous years, the main aim of STS task is to predict the semantic similarity of two sentences in the range 0 to 5 where 0 represents completely different sentences and 5 denotes completely similar sentences [4,5]. In this year Semeval test dataset consists of five different categories with different topics and different textual characteristics like text length or spelling errors: answer-answer, plagiarism, postediting, headlines, and question-question. In SemEval workshop the organizers provide test and training dataset of 2016 along with previous year dataset. Participants can use previous year dataset to train their systems. System quality is determined by calculating the Pearson correlation between the system output values and gold standard values. The system described in this paper explores an alternative approach based on five simple and robust textual similar features. Cosine similarity is used as first feature and second feature simply count the number of words common to the pair of sentences being assessed. The third feature calculates levenshtein ratio needed to transform one sentence into another. METEOR (machine translation metric) is also used to find the similarity score. Finally we are trying to predict the similar score using Gensim [2] toolkit where words and phrases are represented by word2vec [14] language model.

© Springer International Publishing AG 2017
R. Prasath and A. Gelbukh (Eds.): MIKE 2016, LNAI 10089, pp. 144–152, 2017.
DOI: 10.1007/978-3-319-58130-9_14

2   Related Work

Different types of approach have been proposed to predict semantic similarity between sentences based on lexical matching and linguistic analysis [10,11]. For lexical analysis, researchers used edit distance, lexical overlap and largest common sub-string [12] features. Syntactic similarity is another method to find sentence similarity. For syntactic similarity, dependency parses or syntactic trees are used. Knowledge based similarity is mainly based on WordNet. The drawback of knowledge based system is that WordNet is not available for all languages.

On other hand distributional semantics is also used in the field of similarity task. The main idea of distributional semantic is that the meaning of words can be depending in their usage and the context they appear in. The improvement of system can be achieved by stemming, stopword removal, part-of-speech tagging.

3   Dataset

SemEval 2016 organizer provides five types of evaluation dataset in monolingual sub task (i.e. News, Headlines, Plagiarism, Postediting, Answer-Answer and Question-Question).[1] The similarity score of those sentences are calculated by multiple human annotators on a scale from 0 to 5. The details statistics about SemEval monolingual test dataset are described in the Table 1.

Table 1. Statistics of STS-2016 test data

Type	Sentence pair
Answer-Answer	1572
Headlines	1498
Plagiarism	1271
Postediting	3287
Question-Question	1555

4   System Description

Our experiment is divided into three stages. In the first stage, different types of pre-possessing technique are used. Next we calculated semantic similarity score using five types of features. Finally our system trained using two neural networks (i) multilayer feed forward network; and (ii) layered recurrent neural network. Same layered architecture used for both networks and the size of the hidden layer is 10. The details about the feature set are described in the next section. Figure 1 describes the overall architecture of our system.

---

[1] http://alt.qcri.org/semeval2016/task1/index.php?id=data-and-tools.

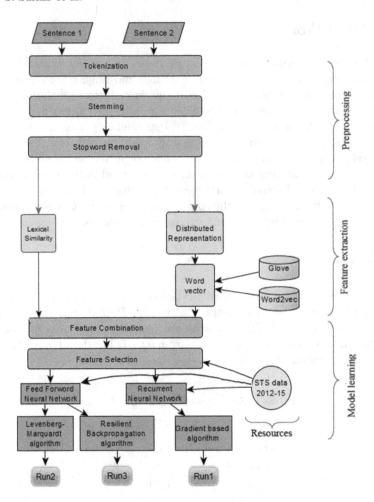

Fig. 1. System description

4.1 Preprocessing

In this section different types of pre-processing techniques are described like tokenization, stopword removal and stemming. The goal of this phase is to reduce inflectional forms of words to a common base form.

(a) **Tokenization**

Sentences can be divided into words only breaking at white-space and punctuation marks. But English language consists of many multi-component words like phrasal verbs. To solve this problem we used NLTK tokenizer. NLTK tokenizer is also required to remove stopword.

(b) **Stemming**

Stemming is an operation in which various forms of words are reduced to a common words. To improve the performance of Information Retrieval system steaming is also used.

(c) **Stop Words**

Stopwords are mainly common words in a language which contain less information. Words like 'a', and 'the' of are appears many times in documents. There is no universal list of stop words. We used NLTK stop word list for our system.[2]

4.2 Features

(a) **Cosine Similarity**

The most commonly used feature for the similarity score is the cosine similarity. In this approach each sentence is represented using vector space model. Cosine similarity is calculated using the dot product by the length of the two vectors. The details description about the cosine similarity is described in Table 2. The cosine similarity between two vectors (S_1, S_2) can be express using this mathematical formula:

$$S = \frac{S_1.S_2}{||S_1||.||S_2||} \tag{1}$$

Table 2. Cosine similarity

Sentence pair	Vector representation	Cosine similarity
Measure the depth of a body of water	Deep = 0, any = 0, measure = 1, the = 1, depth = 1, of = 2, a = 1, body = 1, water = 2, large = 0	0.51639
Any large deep body of water	Measure = 0, the = 0, depth = 0, any = 1, large = 1, deep = 1, body = 1, of = 1, water = 1	

(b) **Unigram matching ratio**

In this approach first total number of similar unigram between two sentences is calculated. Next the similar matching count is divided by the union of all tokens of those sentences. This feature is normalized because similarity score does not depend on the length of sentences. Table 3 describes about this feature where S1 and S2 denotes the sentence pair.

(c) **Levenshtein Ratio**

Levenshtein distance [8] is the difference between two strings. This distance is the minimum number of operation like insertions, deletions or substitutions needed to convert one string to another. Levenshtein distance is similar

---

[2] http://www.nltk.org/book/ch02.html.

Table 3. Unigram matching ratio

Sentence pair	$(S1 \cap S2)/(S1 \cup S2)$
Two green and white trains sitting on the tracks	.83333
Two green and white trains on tracks	

to Hamming distance but Hamming distance is only applicable to the similar length strings. The easiest way to calculate Levenshtein distance using dynamic programming. Levenshtein distance can be used in spell checking where a list of words can be suggest to the user whose levenshtein distance is minimal. The Levenshtein ratio of two strings a, b (of length $|a|$ and $|b|$ respectively) is expressed using Eq. 2. We use the Levenshtein ratio because Levenshtein distance is also depends on the length of the sentences. This feature describes in the Table 4.

$$\text{EditRatio}(a, b) = 1 - \frac{\text{EditDistance}(a, b)}{|a| + |b|} \tag{2}$$

Table 4. Levenshtein ratio

Sentence pair	Levenshtein distance	Levenshtein ratio
TSA drops effort to allow small knives on planes	6	.8958
TSA drops plan to allow small knives on planes		

(d) **Meteor**

Meteor automatic machine translation evaluation system release in the year 2004. Meteor calculates sentence level similarity by aligning them to reference translations and calculating sentence-level similarity scores. To improve the accuracy Meteor uses language specific resources like WordNet and Snowball steamers [6, 7]. For our approach we used Meteor 1.5.[3] Meteor scoring is based on four types of matches (exact, stem, synonym and paraphrase).

(e) **Word2Vec**

In some region similarity between two sentences cannot be decided only using semantic and syntactic analysis. There is a semantic gap between the syntactic structure and the meaning of the sentences because of different vocabulary and language. So we need full knowledge and meaning representation. Using distributional semantic approach the gap between the syntactic meaning and original meaning can be removed. Recently researcher are using Gensim framework where words and phrases are represented using Word2vec [14]

---

[3] https://www.cs.cmu.edu/~alavie/METEOR/README.html.

language model. For our experiment we have used pre-trained word and phrase vectors which are available in Google News dataset [14]. The LSA word-vector mapping model contains 300 dimensional vectors for 3 million words and phrases. Gensim is a Python framework for vector space modeling. We have used Gensim for this experiment, and computed the cosine distance between vectors representing text chunks sentences from SemEval tasks.

5 Results

This section describes the results of our systems for English monolingual STS task of SemEval 2016. System performance measure using Pearson correlation. We used neural network to predict the STS scores. For training process all gold standard training and test data of the year 2012 have used in our task.

In Run 2 We trained our system using Levenberg-Marquardt algorithm and two layer feedforward network with 10 neurons in the hidden layer.[4] In Run 3 similar type of feedforward network is used but trained using Resilient Backpropagation algorithm [9].[5] Similarly in Run 1 our system trained using recurrent neural network [3].[6] However, this performance can be improved by increasing the training dataset and similar type of training and test dataset.

The detail result of the SemEval 2016 monolingual task using Word2vec feature is shown in the Table 5.

Table 5. System performance on SemEval STS-2016 monolingual data using Word2vec

	Run 1	Run 2	Run 3
Answer-Answer	**0.44468**	0.44258	0.44041
Headlines	0.55646	**0.57358**	0.54744
Plagiarism	0.78391	**0.79587**	0.77553
Postediting	0.77594	**0.78888**	0.75682
Question-Question	0.60747	0.61315	**0.62712**

Table 6 describes the result of cosine similarity feature on monolingual test dataset. The results also show that performance on monolingual dataset using only cosine similarity is not suitable for question-question test dataset.

Results in Table 7 show that our approach can achieve better performance except question-question dataset by combining different types of features (i.e. Unigram matching ratio, cosine similarity, lavenshtein ration and METEOR).

---

[4] http://nl.mathworks.com/help/nnet/ref/feedforwardnet.html.

[5] http://nl.mathworks.com/help/nnet/ref/trainrp.html.

[6] http://in.mathworks.com/help/nnet/ug/design-layer-recurrent-neural-networks. html.

Table 6. System performance on SemEval STS-2016 monolingual data using cosine

Corpus	Run 1	Run 2	Run 3
Answer-Answer	**0.42432**	0.41593	0.39188
Headlines	0.52655	**0.52840**	0.51711
Plagiarism	**0.68300**	0.66565	0.66364
Postediting	**0.80030**	0.78705	0.79928
Question-Question	0.13541	0.08708	**0.15116**

Table 7. System performance on SemEval STS-2016 monolingual data using Unigram matching ratio+METEOR+LR+cosine

	Run 1	Run 2	Run 3
Answer-Answer	0.52740	**0.56766**	0.52166
Headlines	0.69000	**0.71222**	0.66787
Plagiarism	0.73626	**0.77102**	0.72979
Postediting	0.75320	**0.77420**	0.74240
Question-Question	0.48372	**0.53835**	0.44357

On the other hand Table 5 shows that word2vec feature gives better result on question-question test dataset. With different type of feature set, we achieved a strong (>0.70%) correlation with human judgments on 3 of the 5 monolingual data set.

6   Compare with Winner Score and Baseline Score

Table 8 describes the comparison between the top ranked system and baseline score with our best result. In English Semantic Textual Similarity (STS) shared task the best result was obtained by Samsung Poland NLP Team.[7] Our System perform well for the postediting dataset. For postediting dataset the difference between the winner result and our result is minimum. However, our system struggles on both of the question-question and answer-answer dataset. Different combination of feature set gives better result on different type of dataset. When we are using word2vec then it gives better result for the question-question, postediting and plagiarism dataset. Similarly the score is high for answer-answer and headline dataset when cosine similarity, unigram matching ratio, levenshtein ratio and METEOR are used. Baseline system is based on unigram matching without stopword, METEOR and levenshtein ratio. In our approach cosine similarity and unigram mating ratio are added to baseline system.

---

[7] http://alt.qcri.org/semeval2016/task1/index.php?id=results.

Table 8. Compare with winner core and baseline score

Corpus	Winner score	Our system		Baseline score
		Best score	Features	
Answer-Answer	0.69235	0.56766	UMR+LR+METEOR	0.48023
Headlines	0.82749	0.71222	UMR+LR+METEOR	0.70749
Plagiarism	0.84138	0.79587	Word2vec	0.76752
Postediting	0.86690	0.80030	Cosine	0.77196
Question-Question	0.74705	0.62712	Word2vec	0.43751

7   Conclusions and Future Work

In this paper we described our experiment on the SemEval-2016 Task 1 monolingual test dataset in Textual Similarity and Question Answering Track. We observed that our system performance vary between different type of dataset. The Pearson correlation of all three runs are 0.8 or above for three test datasets: Headlines, Plagiarism, and Postediting, However the performance of our approach are comparatively lower for Question-question and Answer-answer test datasets. For the future work our aim is to analysis the reason behind the poor performance on answer-answer and question-question dataset. We also plan to include features which are directly based on Wordnet and also try to implement those features to find the similarity for crosslingual dataset.

Acknowledgment. This work presented here is under the Research Project Grant No. YSS/2015/000988 under Science and Engineering Research Board (SERB), Govt. of India. Authors are also acknowledges the Department of Computer Science & Engineering of National Institute of Technology Mizoram, India for providing infrastructural facilities and support.

References

1. Clarke, F., Ekeland, I.: Nonlinear oscillations and boundary-value problems for Hamiltonian systems. Arch. Rat. Mech. Anal. **78**, 315–333 (1982)
2. Rehurek, R., Sojka, P.: Software framework for topic modelling with large corpora. In: Proceedings of LREC 2010 Workshop New Challenges for NLP Frameworks, p. 4550 (2010)
3. Elman, J.L.: Finding structure in time. Cogn. Sci. **14**(2), 179–211 (1990)
4. Agirre, E., Baneab, C., Cardiec, C., Cerd, D., Diabe, M., Gonzalez-Agirrea, A., Guof, W., Lopez-Gazpioa, I., Maritxalara, M., Mihalcea, R., Rigau, G., Uria, L., Wiebe, J.: SemEval- 2015 task 2: semantic textual similarity, English, Spanish and Pilot on interpretability. In: Proceedings of the 9th International Workshop on Semantic Evaluation (SemEval 2015), pp. 252–263 (2015)
5. Agirre, E., Baneab, C., Cer, D., Diab, M., Gonzalez-Agirree, A., Mihalceab, R., Wiebe, J.: SemEval-2016 task 1: Semantic textual similarity - monolingual and cross-lingual evaluation. In: Proceedings of the 10th International Workshop on Semantic Evaluation (SemEval 2016) (2016)

6. Denkowski, M., Lavie, A.: Extending the METEOR machine translation evaluation metric to the phrase level. In: Proceedings of the HLT: The 2010 Annual Conference of the North American Chapter of the Association for Computational Linguistics, Los Angeles, California, pp. 250–253 (2010)

7. Banerjee, S., Lavie, A.: METEOR: an automatic metric for MT evaluation with improved correlation with human judgments. In: Proceedings of the ACL 2005 Workshop on Intrinsic and Extrinsic Evaluation Measures for MT and/or Summarization, Ann Arbor, Michigan, pp. 65–72 (2005)

8. Levenshtein, V.I.: Binary codes capable of correcting deletions, insertions and reversals. Sov. Phys. Dokl. **10**, 707 (1996)

9. Riedmiller, M., Braun, H.: RPROP: a fast adaptive learning algorithm. In: Gelenbe, E. (ed.) International Symposium on Computer and Information Science VII, Antalya, Turkey, pp. 279–286 (1992)

10. Huang, A.: Similarity measures for text document clustering. In: Proceedings of the Sixth New Zealand Computer Science Research Student Conference (NZCSRSC 2008), Christchurch, New Zealand, pp. 49–56 (2010)

11. Aziz, M., Rafi, M.: Sentence based semantic similarity measure for blog-posts digital content. In: 2010 6th International Conference on Multimedia Technology and Its Applications (IDC), pp. 69–74 (2010)

12. Achananuparp, P., Hu, X., Shen, X.: The evaluation of sentence similarity measures data warehousing and knowledge discovery. In: Proceedings of the 10th International Conference, DaWaK 2008, Turin, Italy, 2–5 September 2008, pp. 305–316 (2008)

13. Pennington, J., Socher, R., Manning, C.D.: GloVe: global vectors for word representation. In: Proceedings of the 2014 Conference on Empirical Methods in Natural Language Processing (EMNLP), pp. 1532–1543 (2014)

14. Mikolov, T., Sutskever, I., Chen, K., Corrado, G., Dean, J.: Distributed representations of words and phrases and their compositionality. Adv. Neural Inf. Process. Syst. **26**, 3111–3119 (2013)

Text Mining Models to Predict Brain Deaths Using X-Rays Clinical Notes

António Silva, Filipe Portela[✉], Manuel Filipe Santos, José Machado, and António Abelha

Algoritmi Research Centre, Guimarães, Portugal
a64895@alunos.uminho.pt, {cfp,mfs}@dsi.uminho.pt,
{jmac,abelha}@di.uminho.pt

Abstract. The prediction of events is a task associated to the Data Science area. In the health, this method is extremely useful to predict critical events that may occur in people, or in a specific area. The Text Mining is a technique that consists in retrieving information from text files. In the Medical Field, the Data Mining and Text Mining solutions can help to prevent the occurrence of certain events to a patient. This project involves the use of Text Mining to predict the Brain Death by using the X-Ray clinical notes. This project is creating reliable predictive models with non-structured text. This project was developed using real data provided by Centro Hospitalar do Porto. The results achieved are very good reaching a sensitivity of 98% and a specificity of 88%.

Keywords: X-Rays · Brain death · Text Mining · Predictive medicine

1 Introduction

The X-ray exams are important in determining the clinical condition of a patient. Using X-ray is possible to check if patients have a broken bone, or if there is any significant change in their body. The X-ray is also important to help in determining the existence of Brain Death in a patient. Brain death is one of the worst diagnoses that can be identified in a patient, because it confirms the death of the patient's brain, and the confirmation of their vegetative state. This diagnosis is irreversible and can lead to the patient death. The X-ray clinical notes are stored electronically on the hospital information system. The clinical notes despite being stored in electronic format; they are available in free text. This means that the text written in the notes is not structured, and it can increase the difficulty in obtaining useful information from them. This work aims to create predictive models able to predict in a reliable way, the Brain Death after the execution of an X-ray. The prediction of events is something that begins to be common these days, but the types of data used for the prediction of events are generally unstructured. In this project, the data used to create predictive models were in text format. Besides that, the text was not organized and does not have a structure defined, so this situation difficult the knowledge retrieval from the data. In this analysis, the focus turns to forecast Brain Death based on the X-ray diagnosis. This diagnosis, as mentioned earlier, is a non-structured text format, which could hinder the creation of reliable models, but can also

© Springer International Publishing AG 2017
R. Prasath and A. Gelbukh (Eds.): MIKE 2016, LNAI 10089, pp. 153–163, 2017.
DOI: 10.1007/978-3-319-58130-9_15

open doors to new treatments in medicine by predicting disease and clinical events through clinical notes. To the development of this work, real data (X-Ray notes) provided by Centro Hospitalar do Porto (CHP) were used.

This article has seven sections including the introduction section. The background section presents the context of the problem and the related work that exists in this field. The materials and methods section includes the materials used on the development of this project and the methodologies used in the same. The forth section presents the case study by explain the development of the project following the Cross-Industry Standard Process for Data Mining (CRISP-DM) phases. The achieved results were analyzed in the discussion section. Finally, the conclusion and future work sections, presents the final statements and the possible directions for a future work in this field.

2 Background

2.1 Text Mining

Text Mining (TM) is also known as Text Data Mining, and discovery of textual databases [1]. TM refers to the process of extracting interesting patterns or knowledge from unstructured text documents. Regarding Data Mining, Text Mining is a more complex technique because it involves dealing with text data unstructured and diffuse [2]. Text Mining derives much inspiration and direction of research carried out in Data Mining. There are also other practice areas in the same region of TM as the General Statistics, Machine Learning, Data Base Management, and computational language [3]. This process is achieved by the identification and exploration of patterns of interest in unstructured datasets as books, web pages, e-mails, reports or even product descriptions. It can also be formally defined as the creation of a new non-obvious information (such as patterns, relationships) of a collection of textual documents [1].

2.2 Clinical Decision Support Systems

The Clinical Decision Support Systems (CDSS) are computerized interactive programs aiming to help doctors and other health professionals [4]. The CDSS help in prescription of medications, diagnosis and disease management, to improve services and reduce costs, risks and errors [5]. CDSS can be used to check for drug allergies, comparing costs thereof and different laboratories evaluate the potential interactions between these suggest alternatives, lock duplicate requests, suggesting dosages and provide recommendations. Furthermore, CDSS can provide clinical expertise and best practice standards and guidelines for non-specialists [5]. The CDSS should be integrated with electronic clinical notes systems and computer systems that are connected to other information systems (for example, laboratory, radiology, and billing). The basic components of a CDSS include a basic medical knowledge and an inference engine (usually a set of rules that are derived from experts and evidence-based medicine) and run through medical logic modules based on a language like the Arden syntax, or using an artificial neural network [5].

INTCare is an example of a CDSS and it is working in the Intensive Care Unit of CHP. INTCare uses Knowledge Discovery in Database (KDD) and the agent-based paradigms in order to help decision-making process in medical issues. The INTCare is a system that helps doctors to make decisions by detecting the patient's condition through continuous updates on their health status and applying the forecasting model to predict the occurrence of clinical events. INTCare is divided in four subsystems: Data Acquisition, Knowledge Management, Inference and Interface. INTCare also runs an up-to-date maintenance on the probability of end-of-life decision making. Besides this, INTCare also evaluates the scenarios of the patient's condition, allowing physicians to compare the effects of different medical procedures [6–8].

2.3 Clinical Notes

Professionals of healthcare organizations, such as a hospital, clinic, healthcare center, etc., create electronical clinical notes. This tends to be part of a single information system allowing the storage, collection and handling of medical records to reduce medical error [9]. The electronic clinical notes contain the entire history of the wearer diseases, as well as demographics, current health problems, regular medication, vital signs, immunizations, laboratory data and X-ray reports. The electronic clinical notes automate and create timelines of the patient [9]. The X-Ray used in this project are a type of clinical records that provide the information of the X-Ray performed to the patient. In this work, the data collected was stored in X-Ray reports.

2.4 Brain Death

Brain death is declared when the reflections of the brainstem respiratory drive and motor responses are absent in a norm thermic non-drugged comatose patient, irreversible brain injury with a known mass and do not contribute to metabolic disorders are verified. The determination of brain death in adults has become an integral part of neurological and neurosurgical practice, but it may include any medical specialty [10].

There is a clear difference between severe brain damage and brain death. The physician should understand this difference, because brain death means that support life is useless, and it is the main requirement for the donation of organs for transplant (DOHS). In adults, the main causes of brain death are traumatic brain injuries and hemorrhage [11]. Ethical, religious and philosophical considerations about the definition of death have been addressed in a recent monograph [12].

The interpretation of Computed Tomography (CT), or X-Ray Computed Tomography, is essential for determining the cause of brain death. Typically, the CT reports a patient with a mass brain herniation, multiple hemispheric lesions with edema, or swelling alone. However, such a finding on computed tomography does not eliminate the need for a careful search for confounders. Although that, the CT report can show no sign of problems in patients that had a cardiac arrest earlier and in the patients with fulminant meningitis or encephalitis in an initial stage [13].

There are other confirmatory tests to determine if there is a brain death, like the electroencephalography, Cerebral Angiography, Transcranial Doppler ultrasonography, and a Cerebral scintigraphy [10].

2.5   Related Work

There are some similar projects using Text Mining but applied to other areas of medicine. For example, Pereira et al. [14] made a Text Mining approach to classify the epileptic diagnosis in a child based on the International Classification of Diseases, Ninth Revision (ICD-9). They choose the K-Nearest Neighbor algorithm and the electronic medical records applied as white-box multi classifier approach to classify each instance mapping into the corresponding standard code. Cormack et al. [15], used an agile Text Mining platform to extract document-lever cardiac risk factors in the patient's electronic medical records defined in the i2b2/UT Health 2014 Challenge. They used a data-driven rule-based methodology with the addition of a simple supervised classifier. Jonnagaddala et al. [16] made a project with the aim to extract Framingham risk factors from unstructured electronic health records using clinical text mining. Their project also calculates a 10-year coronary artery disease rick scores in a cohort of diabetic patients. They developed a rule-based system to extract the risk factors.

3   Materials and Methods

3.1   Methods Utilized

In the development of this work several tools were explored: Oracle SQL Developer for extraction of the original Data; Microsoft SQL Server 2014 to select the desired data through a creation of a view with the defined parameters; Microsoft Excel 2014 to prepare the data analyzed by the KNIME Software; KNIME to create the Text Mining workflows and the prediction models.

3.2   Methodologies

Design Science Research (DSR) was the research methodology and the Cross-Industry Standard Process of Data Mining (CRISP-DM) was used to drive the TM work.

Hevner et al. [17] has defined seven guidelines for Design Science in Information Systems Research, and they are the, Design as an artifact, the Relevance of the Problem, the Evaluation of Design, the Contributions Research, the Rigor Search, the Design as a search procedure, and finally The Communication Research.

The CRISP-DM methodology describes the process of Data Mining (in this case, Text Mining) in six stages: the business study, data study, data preparation, modeling, evaluation and implementation. This approach has numerous advantages when applied to DM projects, such as greater speed, lower running costs, greater security and feasibility and viability of projects [18]. The results are in the form of a confusion matrix. The confusion matrix will have the values True Positive (TP), True Negative (TN),

False Positive (FP) and False Negative (FN). Witten et al. [19] defines TP and TN as the correct ratings. A FP is verified when an outcome is incorrectly predicted as positive when its real value is negative. A FN is verified when an outcome is negative, when its real value is positive. To make a more accurate assessment of the models, four metrics were chosen: Accuracy, Error, Sensitivity and Specificity. The Accuracy represents the number of correct predictions (TP and TN). The error is the percentage of wrong model adjustments, the sensitivity evaluates the effectiveness of the classifier to recognize the positive samples and the specificity measures the effectiveness of the classifier to recognize and negative samples [20].

4   Case Study Using Text Mining

4.1   Business Study

The business goal of this work is the prediction of brain death after performing an X-ray by alerting health professionals to take actions to prevent it.

For that, the mining concept was explored using text fields. The TM goal is inducing models able to achieve positive results in the Sensitivity and the Specificity metrics making them available to be used in decision-making process.

4.2   Data Study

This project was developed in conjunction with the Centro Hospitalar do Porto (CHP) – Hospital de Santo António. They provided the necessary data make the prediction models. The data provided were collected from real events happened in the CHP including the deaths recorded in that location, and the X-Rays reports made to patients. Although these data are real, the confidentiality of the latter lies safeguarded.

Fig. 1. Self-Organizing map

Figure 1 presents a set of words associated to patients who died after performing the X-Ray. Each color represents one cluster, and each cluster have their key-terms as can be seen [21]. The project aimed to create patterns of patients that have died after the performing of the X-Ray. This mean that if the patient does not fit in none of this clusters, in all like hood the patient had lived before the X-Ray.

The data were divided into 2 Datasets:

- Dataset with information of deaths, containing the deaths recorded in hospital since the December 6, 2005 up to March 1, 2016, in a total of 14.528 records.
- Dataset with X-ray reports containing the information of X-Rays (XR) performed at the hospital since the October 12, 2009 until the day May 17, 2016 in a total of 147.973 records.

The data were stored in an Oracle SQL database, then the files were extracted to a spreadsheet. Thus, there was an analysis of the dataset in order to identify the most relevant and non-relevant fields. The most relevant data of the Death dataset are described below:

- OBIT_EMER - Identifies if the death occurred;
- SEQUENTIAL_NUM – number of arrival;
- PROCESS_NUM – patient identification;

The most relevant data of the XR dataset are described below:

- EXAMNUM – exam number;
- PROCESS – patient identification;
- TITLE – title of the exam performed;
- DESCRIPTION – exam clinical notes in free text.

From these two datasets there is a common attribute (PROCESS) which allow a connection between the two tables. When the connection was performed, a new table was created containing only the XR data from death patients. First, the two datasets were imported as tables to the Microsoft SQL Server after that, two views were created to select the specific data. The objective of the two views was to divide the X-Ray data in 2 groups, and cleaning the data that had errors such as null values in the DESCRIPTION and in the PROCESS column. This view makes a selection of the cranioencephalic X-Rays (X-Ray Computed Tomography) records. With this selection there are two groups: lived patients after the X-Ray and the group of patients that died after the X-Ray. As result, a new view called Live Reports was created containing 30.904 lines, and another view namely Death Reports containing 4.899 lines. Finally, the views were extracted into a spreadsheet file. After this process, the number of X-Rays associated to death patients represents only 13.68% of all Cranioencephalic X-Rays performed.

4.3   Data Preparation

In the data preparation phase some issues were found. The data contained many lines and the testing machine was not able to read all the data, so a data sample containing the X-rays performed to the patients between the year 2009 and 2010 was created. The result of this filtering was the creation of two spreadsheets with the name of DeathReports20092010 that have the information of the X-Rays of patients that have died after the exam, and the spreadsheet LifeReports20092010 containing the X-Rays performed to the patients who did not die in that years. The files contained a considerable disparity between the DeathReports20092010 file and LifeReports20092010 file. The DeathReports20092010 file had 1094 lines, and the LifeReports20092010 file contained 4434 lines. The solution was to copy and paste the existing lines in DeathReports20092010 file in order to achieve a similar number to the LifeReports20092010 file keeping the data integrity. This method is called oversampling, and the result was the creation of a new file, DeathReports20092010Oversampling, which contains 4373 lines, which are already a number of lines similar to LifeReports20092010 file, which allowed making a fairer analysis of the data, but the file that does not contain oversampling will be analyzed as well. These models want to be always with the greatest possible accuracy and a threshold were defined to the four metrics already mentioned. The criteria were defined having as base Data Mining good practices and the healthcare professionals' knowledge. A model to be select should meet the following threshold: The sensitivity and accuracy upper than 85% and the specificity upper than 75%.

4.4 Modeling

In this phase, TM models were induced using four distinct configurations (Table 1): Decision Tree, K-Nearest Neighbor, and Decision Tree with Cross Validation and K-Nearest Neighbor with Cross Validation. As sampling methods two approaches were used: with or without oversampling. Each algorithm had a Holdout sampling where 30% of the data were used for test, and 70% of the data was the training set. The following table show more details about the algorithms used in the project.

Three different scenarios were created. These scenarios include the use of a POS tagger, and the use of a Tag Filter or lack of these methods:

- S1 – {POS Tagger, Tag Filter};
- S2 – {POS Tagger};
- S3 – {Without POS Tagger and without Tag Filter}.

The Text Mining models were inducted in the target approaches. Each TM model is defined by the following equation:

$$TMM_m = TMT_y \times A \times T_s \times S_i \qquad (1)$$

The TMT_y refers to the Text Mining technique, the A is the target approach, the T_s is the Sampling method and the S_i refers to the Scenario. Twenty-four models were induced (4 mining techniques × 1 target approach × 2 sampling method × 3 scenarios).

4.5 Evaluation

The achieved results are present in Table 2, where the Accuracy (Acc.), Specificity (Spec.), Sensitivity (Sens.), and Error (Er.) are expressed by percentage. Table 2 presents

Text Mining Models to Predict Brain Deaths Using X-Rays 7 the best result (which achieved the threshold) by scenario, the respective algorithm and if the model used or not oversampling (OS). Table 2 presents the best results using or not oversampling. The values of Sensitivity, Specificity, Accuracy and Error are defined by the following equations:

Table 1. Details of the algorithms used

	Parameter	Value
Decision tree	Sample	30%
	Number of seeds	Random
	Quality measure	Gini Index
	Pruning method	No pruning
	Min number records per node	2
	Number threads	4
K-nearest neighbour	Sample	30%
	Number of seeds	Random
	Number of neighbours to consider (k)	5
	Weight neighbours by distance	True
Decision tree with cross validation	Sample	30%
	Number of seeds	1441778237149
	Quality measure	Gini Index
	Pruning method	No pruning
	Min number records per node	2
	Number threads	8
	Number of validations	10
K-nearest neighbour with cross validation	Sample	30%
	Number of seeds	1441778237149
	Number of neighbours to consider (k)	3
	Weight neighbours by distance	True
	Number of validations	10

Table 2. Best models results

Algorithm	OS	Scenario	Sens.	Spec.	Acc.	Er.
KNN	No	1	3.05	**98.12**	79.31	20.69
DT	No	1	28.05	**85.49**	74.13	25.87
KNN (CV)	No	2	8.23	**96.53**	79.06	20.94
DT (CV)	No	2	24.70	**85.02**	73.09	26.91
DT (CV)	No	3	27.63	**84.46**	73.22	26.78
KNN	Yes	1	**97.33**	75.34	**86.26**	**13.74**
DT (CV)	Yes	1	**97.96**	79.36	**88.60**	**11.40**
DT (CV)	Yes	2	**98.92**	77.55	**88.16**	**11.83**
DT (CV)	Yes	3	**98.81**	78.77	**88.72**	**11.28**

$$Sensitivity = 100 \times (TP/(TP + FN))$$

$$Specificity = 100 \times (TN/(TN + FP))$$

$$Accuracy = 100 \times (\frac{TP + TN}{TP + FP + TN + FN})$$

$$Error = 100 - Accuracy$$

5   Discussion

The best results were achieved using oversampling. The datasets without oversampling achieved good results in the specificity, with the best result reaching 98%, but failed to have satisfactory results in de Sensitivity, where the best result was 30%. This means that the models are very good to predict survivors and not to anticipate a possible brain death. The Decision Tree without cross validation achieved good results in the Specificity, where the best result achieved is 76.7%, and sensitivity fields, where the best result achieved 91%, but failed to meet the standard values. The K-Nearest Neighbor had similar results when compared the use of cross validation or the lack off, with very good results in the Sensitivity field, with 98%, but the other results were not satisfactory. The algorithm that had the best results, was bay far, the Decision Tree with cross validation, where the all the results meets the minimum values defined. In this algorithm, the best scenario is the scenario 3 (Without POS Tagger and without Tag Filter) with the following values (98.81% in Sensitivity, 78.77% in the Specificity and 88.72% in the Accuracy. In addition to complying the standard values, this result exceeds largely the Sensitivity metric. These values were achieved with the model that does not use any kind of enrichment of words in the workflow, which is interesting and can open new possibilities to research in this area.

6   Conclusion and Future Work

In this project, the major objective was achieved with a creation of predictive models that meet the thresholder earlier defined. The results are very promising in predicting the death (1). The sensitivity is around 99% and the error percentage a little upper than 10%. These models are prepared to be included in an Intelligent Decision Support System in order to be used by the healthcare professionals. Now, the models can indicate the percentage of a patient have a brain death. This knowledge is very useful in order to prevent a possible Brain Death in a patient by applying preventive medicine. Besides that, this project also attests the possibility of explore clinical notes using text mining to achieve new knowledge. In this area some other approach will be explored using other methods, such like the use of a dictionary tagger with specific terms, and this can help to make a more focused prediction of data.

The future work of this project can focus in analyzing other type of clinical notes to verify if it is possible to predict different type of diseases, or events. This can be very useful for the patients because can allow to the health-care professionals to prevent a critical event in a patient, and so improve their life quality.

Acknowledgments. This work has been supported by Compete: POCI-01-0145-FEDER-007043 and FCT - Fundação para a Ciência e Tecnologia within the Project Scope UID/CEC/00319/2013.

References

1. Hearst, M.A.: Untangling text data mining. In: Proceedings of 37th Annual Meeting of the Association for Computational Linguistics on Computational Linguistics, pp. 3–10 (1999). doi:10.3115/1034678.1034679
2. Tan, A.-H.: Text mining: the state of the art and the challenges. In: Proceedings of the PAKDD 1999 Workshop on Knowledge Disocovery from Advanced Databases, vol. 8, pp. 65–70 (1999). doi:10.1.1.132.6973
3. Truyens, M., Van Eecke, P.: Legal aspects of text mining. Comput. Law Secur. Rev. **30**(2), 153–170 (2014). doi:10.1016/j.clsr.2014.01.009
4. Gamberger, D., Prcela, M., Jovi, A., Šmuc, T., Candelieri, A., Conforti, D., Guido, R.: Medical knowledge representation within Heartfaid platform. In: Proceedings of Biostec 2008 International Joint Conference on Biomedical Engineering Systems and Technologies, pp. 205–217 (2008). doi:10.5220/0001045203070314
5. El-sappagh, S.H.: A distributed clinical decision support system architecture. J. King Saud Univ. – Comput. Inf. Sci. **26**(1), 69–78 (2014). doi:10.1016/j.jksuci.2013.03.005
6. Portela, F., Pinto, F., Santos, M.F.: Data mining predictive models for pervasive intelligent decision support in intensive care. In: KMIS 2012 - Knowledge Management and Information Sharing, pp. 81–88. SciTePress, Barcelona (2012)
7. Portela, F., Santos, M.F., Machado, J., Abelha, A., Silva, Á.: Pervasive and intelligent decision support in critical health care using ensembles. In: Bursa, M., Khuri, S., Renda, M.E. (eds.) ITBAM 2013. LNCS, vol. 8060, pp. 1–16. Springer, Heidelberg (2013). doi: 10.1007/978-3-642-40093-3_1. ISBN 978-3-642-40093-3
8. Veloso, R., Portela, F., Santos, M.F., Silva, Á., Rua, F., Abelha, A., Machado, J.: Using domain knowledge to improve intelligent decision support in intensive medicine - a study of bacteriological infections. In: ICAART 2015 - 7th International Conference on Agents and Artificial Intelligence, pp 582 – 587. SciTePress, Lisbon (2015). ISBN:978-989-758-074-1
9. Alanazi, H.O., Jalab, H.A., Alam, G.M., Zaidan, B.B., Zaidan, A.A.: Securing electronic medical records transmissions over unsecured communications: an overview for better medical governance. J. Med. Plants Res. **4**(19), 2059–2074 (2010). doi:10.5897/JMPR10.325
10. Wijdicks, E.F.M.: Brain death worldwide (2007). doi:10.1212/WNL.58.1.20
11. Wijdicks, F.M.E.: Determining brain death in adults, May 1995. doi:10.1212/WNL.45.5.1003
12. Brock, D.: The role of the public in public policy on the definition of death. In: Younger, S.J., Arnold, R.M., Schapiro, R. (eds.) The Definition of Death: Contemporany Controversies, pp. 293–307. Hopkins University Press, Baltimore (1999)
13. Greer, D.M., Varelas, P.N., Haque, S., Wijdicks, E.F.M.: Variability of brain death determination guidelines in leading US neurologic institutions. Neurology **70**(4), 284–289 (2008). doi:10.1212/01.wnl.0000296278.59487.c2
14. Pereira, L., Rijo, R., Silva, C., Agostinho, M.: ICD9-based text mining approach to children epilepsy classification. Procedia Technol. **9**, 1351–1360 (2013). doi:10.1016/j.protcy.2013.12.152

15. Cormack, J., Nath, C., Milward, D., Raja, K., Jonnalagadda, S.R.: Agile text mining for the i2b2/UTHealth cardiac risk factors challenge. J. Biomed. Inform. **58**, S120–S127 (2014). doi: 10.1016/j.jbi.2015.06.030
16. Jonnagaddala, J., Liaw, S.-T., Ray, P., Kumar, M., Chang, N.-W., Dai, H.-J.: Coronary artery disease risk assessment from unstructured electronic health records using text mining. J. Biomed. Inform. **58**, S203–S210 (2015). doi:10.1016/j.jbi.2015.08.003
17. Hevner, A., March, S., Park, J.: Design science in information systems research. MIS Q. **28**(1), 75–105 (2004). doi:10.2307/25148625
18. Santos, M.F., Azevedo, C.S.: Data Mining: Descoberta de Conhecimento em Bases de Dados. FCA - Editora de informática, Lda, Lisbon (2005)
19. Witten, I., Frank, E., Hall, M.: Data Mining: Practical Machine Learning Tools and Techniques. Morgan Kaufmann, Massachusetts (2011)
20. Halkidi, M., Vazirgiannis, M.: Quality assessment approaches in data mining. In: Maimon, O., Rokach, L. (eds.) Data Mining and Knowledge Discovery Handbook, pp. 613–640. Springer, Heidelberg (2005). doi:10.1007/978-0-387-09823-4_31
21. Silva, A., Portela, F., Santos, M.F., Machado, J., Abelha, A.: Towards of automatically detecting brain death patterns through text mining. In: IEEE Conference on Business Informatics - ISA'HEALTH@CBI 2016 - Intelligent Systems and Applications in Healthcare Workshop (2016)

Inferring the Repetitive Behaviour from Event Logs for Process Mining Discovery

Tonatiuh Tapia-Flores[(✉)] and Ernesto López-Mellado

CINVESTAV Unidad Guadalajara,
Av. del Bosque 1145, Col. El Bajio, 45015 Zapopan, Mexico
{ttapia,elopez}@gdl.cinvestav.mx

Abstract. This paper addresses the problem of discovering a sound Workflow net (WFN) from event traces representing the behavior of a discrete event process. A novel and efficient method for inferring the repetitive behaviour in a workflow log is proposed. It is based on an iterative search and filtering of cycles computed in each trace; a graph of causal relations is built for each cycle, which helps to find the supports of the t-invariants of an extended WFN. The t-invariants are used for determining causal and concurrent relations between events, allowing building the WFN efficiently in a complete discovery technique.

Keywords: Process mining · Petri nets discovery · t-invariants

1 Introduction

The synthesis of formal models from external observation of systems behaviour is nowadays an interesting and challenging approach for reverse engineering of discrete event processes. Although the problem is relatively recent, it deserves the attention of several research groups in the fields of discrete event systems (DES) and workflow management systems (WMS).

Pioneer works on the matter, named language learning techniques, have been proposed in the field of computer sciences. The aim was to build fine representations (finite automata, grammars) of languages from samples of accepted words [4,12].

In the field of DES, where the problem is named identification, several approaches have been proposed for building models representing the observed behaviour of automated processes. The incremental approach proposed in [15,16], allows building safe interpreted Petri net (PN) models from a continuous stream of system's outputs. In [11], a method based on the statement and solution of an integer linear programming problem is proposed; it allows building PN from a set of sequences of events. Extensions of this method have been proposed in [6,8]. In [13] a method for deriving finite automaton from sequences of inputs and outputs is presented; it is applied to fault detection of manufacturing processes; an extension to this method allows obtaining distributed

© Springer International Publishing AG 2017
R. Prasath and A. Gelbukh (Eds.): MIKE 2016, LNAI 10089, pp. 164–173, 2017.
DOI: 10.1007/978-3-319-58130-9_16

system models is presented in [17]. In [9] input-output identification of automated manufacturing process is addressed; an interpreted PN is obtained from
a set of sequences of input-output vectors collected from the controller during
the system cyclic operation. The method is extended for dealing with complex
behaviors in [18]. More complete reviews on DES identification can be found
in [5,10].

In WMS the analogous problem is named process mining discovery; the system observation is given as a set of sequences from a finite alphabet of tasks,
representing execution logs of business processes. A first proposal is reported in
[3], in which a finite automaton, called conformal graph is obtained. In [7] it
is proposed a probabilistic approach to find the concurrent and direct relations
between tasks. The input of the method is a sequence of events representing the
activities that have occurred in a workflow management system; the obtained
model is graph similar to a PN. An approach based on conjoint occurrent classes
is presented in [19]. In [1] a mining method called algorithm alpha is presented. In
this method a workflow tasks log composed by several traces is recorded sequentially and processed yielding a subclass of PN called workflow net. Numerous
publications present extensions of this algorithm, namely [20–22], Other related
works can be found in [2].

In the present paper, a novel method that deduces the support of t-invariants
from a workflow log is proposed for addressing the process mining discovery problem. The technique allows finding the set of tasks corresponding to repetitive
components from a large log of sampled tasks traces, which represent the repetitive execution of cases in complex business processes exhibiting concurrency and
iterations.

The method is based on determining, from a log, causal and concurrency relations between tasks and the computing of the t-invariants of the PN to discover.
The obtained invariants allow for instance, determining the initial structure of
a PN by finding structured blocks of the original process. All the algorithms
issued from this method, which have polynomial time complexity, have been
implemented and tested with artificial logs generated with know WFN inspired
from reported models in literature.

The paper is organised as follows. In Sect. 2, the basic notions on PN and
WFN are recalled. Section 3 states the addressed problem. Section 4 presents a
technique for determining the t-invariants from the log. Finally, Sect. 5 discusses
the future work and conclude the paper.

2   Background

This section presents the basic concepts and notations used in this paper.

Definition 1. *An ordinary Petri Net structure G is a bipartite digraph rep-
resented by the 4-tuple $G = (P, T, I, O)$ where: $P = \{p_1, p_2, ..., p_{|P|}\}$ and
$T = \{t_1, t_2, ..., t_{|T|}\}$ are finite sets of vertices named places and transitions respectively; $I : P \times T \rightarrow \{0, 1\}$ ($O : T \times P \rightarrow \{0, 1\}$) is a function representing the
arcs going from places to transitions (from transitions to places).*

Incidence matrix: *the incidence matrix of G is $C = C^+ - C^-$, where $C^- = [c_{i,j}], c_{i,j} = I(p_i, t_j)$; and $C^+ = [c_{i,j}], c_{i,j} = O(p_i, t_j)$ are the pre and post matrices respectively.*

Marking function: $M : P \to \mathbb{N}$ *represents the number of tokens residing inside each place; it is usually expressed as a $|P|$-entry vector.*

Petri Net system: (PN) *is the pair $N = (G, M_0)$, where G is a PN structure and M_0 is an initial marking. In a PN system, a transition t_j is enabled at marking M_k if $\forall p_i \in P, M_k(p_i) \geq I(p_i, t_j)$; an enabled transition t_j can be fired reaching a new marking M_{k+1}. The reachability set of a PN is the set of all possible reachable markings from M0 firing only enabled transitions; this set is denoted by $R(G, M_0)$. For any $t_j \in T$, $\bullet t_j = \{p_i | I(p_i, t_j) = 1\}$, and $t_j \bullet = \{p_i | O(t_j, p_i) = 1\}$, similarly for any $p_i \in P$, $\bullet p_i = \{t_j | O(t_j, p_i) = 1\}$, and $p_i \bullet = \{t_j | I(p_i, t_j) = 1\}$.*

t-invariant: *A t-invariant Y_i of a PN is a positive integer solution to the equation $CY_i = 0$. The support of Y_i denoted as $\langle Y_i \rangle$ is the set of transitions whose corresponding elements in Y_i are positive. Y_i is minimal if its support is not included in the support of other $t-invariant$. A t-component $G(Y_i)$ is a subnet of PN induced by a $\langle Y_i \rangle : G(Y_i) = (P_i, T_i, I_i, O_i)$, where $P_i = \bullet \langle Y_i \rangle \langle Y_i \rangle \bullet, T_i = \langle Y_i \rangle, I_i = P_i \times T_i \cap I$, and $O_i = P_i \times T_i \cap O$.*

A PN system is 1-bounded or safe iff, for any $M_i \in R(G, M_0)$ and any $p \in P, M_i(p) \leq 1$. A PN system is live iff, for every reachable marking $Mi \in R(G, M_0)$ and $t \in T$ there is a reachable marking $M_k \in R(G, M_0)$ such that t is enabled in M_k.

Definition 2. *A WorkFlow net (WFN) N is a subclass of PN owning the following properties [2]: (1) it has two special places: i and o. Place i is a source place: $\bullet i = \emptyset$, and place o is a sink place: $o \bullet = \emptyset$. (2) If a transition τ is added to PN connecting place o to i, then the resulting PN (called **Extended WFN**) is strongly connected.*

Sound WFN: *A WFN N is said to be sound iff (N, M_0) is safe and for any marking $M_i \in R(N, M_0)$, $o \in M_i \to M_i = [o]$, N does not contain dead transitions.*

A consecutive activities sequence: $\langle x, d, e, a, b, y \rangle$ of a process executions that brings the system from initial state into the final state, corresponds to a *trace*; a *workflow log* is a multiset of traces. The set of traces that can be produced by a model N, is the language of N, is denoted by $\mathcal{L}(N)$. An example of extended workflow net is shown in Fig. 1.

Definition 3 *(Relations between activities). Let λ be a log over T. and let $a, b \in T$:*
- $a \to_\lambda b$, *iff there is a trace $\sigma = \langle t_1, t_2, ...t_n \rangle \in \lambda$ and $1 \leq i < n - 1$, such that $t_i = a$, $t_{i+1} = b$,*
  - $a ||_\lambda b$ *iif $a \to_\lambda b$ and $b \to_\lambda a$.*
  - $first(\sigma) = t_1$, $last(\sigma) = t_n$, *if $n \geq 1$*

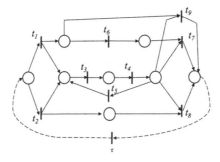

Fig. 1. Extended workflow net N.

The ordering relations has been used in [1,3] as an abstraction of the behavior described by a log or model.

3   Problem Formulation and Method Overview

3.1   Model Discovery

The problem of Petri net model discovery in the context of workflow management systems is formulated.

Definition 4. *Given a workflow log* $\lambda = \{\sigma_1, \sigma_2, \ldots \sigma_j\}$ *generated by a sound WFN, the process mining discovery problem consist of synthesizing a safe PN structure using only transitions in* T, *which reproduces the observed log. The number of places is unknown.*

3.2   Finding the Repetitive Components

Definition 5. *Given a workflow log* $\lambda = \{\sigma_1, \sigma_2, \ldots \sigma_j\}$ *generated by a sound WFN, the problem of finding the repetitive components consists of deducing sets of tasks* $Y_i \subseteq T$, *which correspond to the supports t-invariants of the original WFN using only the information in* λ.

Example 1. Consider the log $\lambda = \{\sigma_1, \sigma_2, \sigma_3, \sigma_4, \sigma_5, \sigma_6, \sigma_7\}$ composed by following tasks traces as the result of the execution of the WFN in Fig. 1. $\sigma_1 = \langle t_1, t_6, t_3, t_4, t_7 \rangle$; $\sigma_2 = \langle t_1, t_3, t_6, t_4, t_5, t_3, t_4, t_7 \rangle$; $\sigma_3 = \langle t_2, t_3, t_4, t_5, t_3, t_4, t_8 \rangle$; $\sigma_4 = \langle t_2, t_3, t_4, t_8 \rangle$; $\sigma_5 = \langle t_1, t_3, t_4, t_5, t_6, t_3, t_4, t_7 \rangle$; $\sigma_6 = \langle t_1, t_3, t_4, t_5, t_3, t_4, t_6, t_7 \rangle$; $\sigma_7 = \langle t_1, t_3, t_4, t_9 \rangle$. The repetitive components of this log are $\langle Y_1 \rangle = \{t_1, t_3, t_4, t_6, t_7, \tau\}$, $\langle Y_2 \rangle = \{t_1, t_3, t_4, t_8, \tau\}$, $\langle Y_3 \rangle = \{t_3, t_4, t_5\}$, $\langle Y_4 \rangle = \{t_1, t_3, t_4, t_9, \tau\}$, which correspond to the t-invariants in the extended workflow net in Fig. 1.

Assumptions. It is assumed that the process is well behaved, i.e. there are no faults, deadlocks, or overflows during the observation period. This is a realistic assumption since the processes whose models have to be discovered are supposed to be in operation, although the model is currently unknown or ill known. Additionally, we consider the that the log represents only the process behavior and there are not duplicate tasks, (i.e. a deadlock-free, 1-bounded workflow net including a transition τ relying the place o to place i) to be discovered. In contrast to the work in [14] the use of the silent activity τ is outside the scope of this article.

3.3   Overview of the Method

The aim of the proposed method is to obtain the repetitive behavior of an unknown or ill known process, the proposed method deduce sets of tasks that represents the supports of t-invariants of the original process (a extended sound workflow net). It focuses on the iterative detection and filtering of cycles in the traces; then a graph of causal relations between tasks is made for each detected cycle, which helps to determine the minimal supports of t-invariants.

4   Computing T-Invariants

In most of the workflow processes the tasks are executed sequentially by cases (tasks traces); if for all traces, their tasks appear only once, the t-invariants are trivially determined because there are no repetitive nested behaviors in the process. However, there are processes that have repetitive sub processes inside, as those modeled by the WFN in Fig. 1; for such processes, deriving the minimum support of t-invariants is not trivial. This section presents an algorithm to compute the minimal supports of t-invariants from a workflow log that have tasks traces describing repetitive behavior.

The computed t-invariants are valid only for the PN model to be built since the only available information is the log. Thus, the technique presented herein pursues to find out the t-invariants of the net that reproduces the observed behavior. Several notions used for defining the t-invariants computation technique are given below.

Definition 6. *A tasks trace that includes t_i whose number of occurrences ($\#t_i$) is greater than one, contains sub-sequences representing a repetitive behavior. We will call **cyc** to the subsequences starting with t_i until before the next occurrence of t_i in σ. A **cyc** may contain nested cycs; if for every t_i in a **cyc**, $\#t_i = 1$, then it is called elementary **cyc**, denoted as **cyc_e**.*

Example 2. Some tasks traces of Example 1 have *cyc* within them; such is the case of $\sigma_2 = \langle t_1, t_3, t_6, t_4, t_5, t_3, t_4, t_7 \rangle$, whose *cyc* = $\{t_3, t_6, t_4, t_5\}$; furthermore it is elementary (**cyc_e**), while σ_1 and σ_4 do not contain a *cyc*.

Proposition 1. *The set of tasks in every trace σ are the support of a t-invariant of the extended WFN. The t-invariant is minimal if $\forall a \in \sigma, \#a = 1$.*

Proof. As stated in Definition 2, the extended WFN is strongly connected. Thus the transitions in σ, whose first and last tasks of σ belong to i. and $.o$ respectively, together with τ can be fired repeatedly. When $\forall a \in \sigma, \#a = 1$, then σ does not have nested cyc therefore the tasks in σ are the support of a minimal t-invariant.   □

Proposition 2. *If $\sigma \in \lambda$ includes cycs, then it must contain tasks belonging to two or more t-invariants.*

Proof. The repeated tasks in σ_i are fired during the execution of nested t-invariants, which do not need the firing of τ to be repeatedly executed. Thus some transitions are executed sometimes in a nested cycle, and sometimes in the cycle including τ.   □

The strategy of the proposed algorithm includes the pruning of interleaving tasks in an event trace in order to separate them by supports of minimal t-invariants. The procedure to obtain the t-invariants operates recursively on every tasks trace σ_i from the most external cyc in σ_i to smaller nested cyc.

Definition 7. *The causality graph G_r of a cyc_e, describes the relations between tasks. $G_r(cyc_e) = (V, E)$; where $V = \{a | a \in cyc_e\}$ and $E = \{(a, b) \in V \times V | a \to b \wedge b \not\to a\}$ (Fig. 2).*

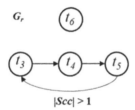

Fig. 2. G_r graph of the cyc_e in Example 2.

Definition 8. *The function $Scc(G_r)$ returns a set of strongly connected components $\{G^1_{sc}(V_1, E_1), \ldots G^2_{sc}(V_2, E_2), \ldots G^n_{sc}(V_n, E_n)\}$ in G_r.*

Proposition 3. *Let $G^i_{sc}(V_i, E_i)$ be a strongly connected component in a Gr, then V_i is a support of a minimal t-invariant of the final model if $|V_i| > 1$.*

Proof (contra position). Suppose that tasks in V_i are not the support of a t-invariant, then there exist at least a $t_k \notin V_i$ that must be fired for allowing the repetitive firing of transitions in G_r; thus there are not cycles in V_i, consequently, it is not strongly connected.   □

Now, let us define two useful operators allowing handling tasks traces.

Definition 9. *The operator clear(s, A), where s is a sequence, and A is a set ($|A| > 0$), yields a new sequence in which the occurrences of $t_i \in A$ in s are deleted. The operation replace(r, s, t), where r, s, and t are sequences ($|r|, |s|, |t| > 0$), yields a sequence in which the first occurrence of s in r is replaced by t.*

In the operator $clear(s, A)$, if none of the elements in A is present in s, the result of this operator is the sequence s; if A contains exactly the elements present in s the operator will return an empty sequence. The $replace(r, s, t)$ operator returns r if s is not a subsequence of r and returns t if r is equal to s; when $t = \{\emptyset\}$, the first occurrence of substring s is deleted form r.

The result of applying the above definitions to $\sigma_2 = \langle t_1, t_3, t_6, t_4, t_5, t_3, t_4, t_7 \rangle$ of the previous example, is $clear(cyc_e, A) = t_6$ where $cyc_e = \{t_3, t_6, t_4, t_5\}$, $A = \{t_3, t_4, t_5\}$. $replace(\sigma_2, cyc_e, t_6) = \{t_1, t_6, t_3, t_4, t_7\}$.

Now, a useful procedure for extracting elementary cycles of traces is given. It explores recursively nested cyc and returns one elementary a cyc_e if there exists; otherwise returns an empty set.

Algorithm 1. $e\text{-}cycle(\sigma)$

Input: Tasks trace σ.
Output: cyc_e.

```
 1: cyc ← ∅
 2: for ∀t_i ∈ σ do
 3:    if #t_i > 1 then
 4:       i ← index of t_i ∈ σ
 5:       j ← index of next t_i ∈ σ
 6:       cyc ← subseq(i, j − 1)
 7:       e-cycle(cyc)
 8:    end if
 9: end for
10: cyc_e ← cyc
```

Now the procedure for obtaining the minimal t-invariants supports is given. $Sym(\sigma_i)$ returns the set of transitions included in σ_i.

The supports of t-invariants resulting from the application of the above algorithm to the tasks traces in Example 1 are $\langle Y_1 \rangle = \{t_1, t_3, t_4, t_6, t_7, \tau\}$, $\langle Y_2 \rangle = \{t_1, t_3, t_4, t_8, \tau\}$, $\langle Y_3 \rangle = \{t_3, t_4, t_5\}$, $\langle Y_4 \rangle = \{t_1, t_3, t_4, t_9, \tau\}$. Notice that for $\langle Y_1 \rangle$, $\langle Y_2 \rangle$ and $\langle Y_4 \rangle$ the task τ is added, because such invariants have transitions in $i^\bullet$ and $^\bullet o$, whilst for $\langle Y_3 \rangle$, τ is not necessary because is a support of a nested t-invariant.

5   Concluding Remark

The idea of inferring the repetitive behavior from a workflow log is to get more information about the process, in order to improve the process mining discovery

Algorithm 2. *Getting the minimal t-invariants supports*

Input: λ.

Output: $Y(\lambda)$.

1: $Y(\lambda) \leftarrow \emptyset$ and $A \leftarrow \emptyset$
2: **for** $\forall \sigma_i \in \lambda$ **do**
3:    **if** $\forall t_x \in \sigma_i, \#t_x = 1$ and $\sigma_i \notin Y(\lambda)$ **then**
4:       $Y(\lambda) \leftarrow Y(\lambda) \cup Sym(\sigma_i)$
5:    **else**
6:       **repeat**
7:          $cyc_e \leftarrow e\text{-}cycle(\sigma_i), G \leftarrow Scc(Gr(cyc_e))$
8:          $\forall G^i \in G$
9:          **if** $|V_i| > 1$ and $V_i \notin Y(\lambda)$ **then**
10:            $Y(\lambda) \leftarrow Y(\lambda) \cup V_i$
11:            $A \leftarrow A \cup V_i$
12:          **end if**
13:          $\sigma_i \leftarrow replace(\sigma_i, cyc_e, clear(cyc_e, A))$
14:       **until** $cyc_e \neq \emptyset$
15:    **end if**
16:    **if** $\sigma_i \notin Y(\lambda)$ **then**
17:       $Y(\lambda) \leftarrow Y(\lambda) \cup Sym(\sigma_i)$
18:    **end if**
19: **end for**

techniques. Given that the log is generated by a sound workflow net, the repetitive behavior found, is actually the t-invariants set of the extended net, which provides a better understanding of the real process.

In this paper a novel technique for inferring the repetitive behaviour in a workflow log has been presented. It is based on an iterative search and filtering of cycles found in each trace; for this purpose, a graph of causal relations is built for each cycle, which helps to find the supports of the t-invariants of the original WFN. The proposed notions and algorithms support a complete method of process mining discovery, which is not described herein due to space constraints. The algorithms run in polynomial-time, which allows processing large logs efficiently including arbitrary long traces; they have been implemented and tested on artificial logs generated for known sound WFN of diverse structural complexity.

A promising research direction is the use of t-invariants to improve existing discovery techniques by taking advantages of the knowledge of the t-invariants, namely

- Using the t-invariants, to separate the process into structural blocks, one for each t-invariant.
- Given that a t-invariant in a Petri net must be strongly connected, we can find more complex behaviors not exhibited in the log i.e, implicit dependencies [22], by assuring this connectivity.
- Finding concurrency, since two tasks in the same t-invariant which are not directly linked must be concurrent.

- Deduce a partial order among the tasks in a t-invariant, which leads to easy substructures task-place-task.

5.1   Conclusion

In this work a technique for inferring the repetitive behavior in a workflow log is presented. It is based on an iterative search and filtering of cycles found in each trace, a graph of causal relations is built for each cycle which helps to find the supports of the t-invariants of the original WFN. Proposed notions and algorithms seem to be a promising approach for addressing the problem of process mining discovery. The algorithms used for this propose run in polynomial time, which allows processing large logs including arbitrary long traces efficiently.

References

1. Van der Aalst, W., Weijters, T., Maruster, L.: Workflow mining: discovering process models from event logs. IEEE Trans. Knowl. Data Eng. **16**(9), 1128–1142 (2004)
2. van der Aalst, W.M.P.: Process Mining: Discovery Conformance and Enhancement of Business Processes, 1st edn. Springer, Heidelberg (2011)
3. Agrawal, R., Gunopulos, D., Leymann, F.: Mining process models from workflow logs. In: Schek, H.-J., Alonso, G., Saltor, F., Ramos, I. (eds.) EDBT 1998. LNCS, vol. 1377, pp. 467–483. Springer, Heidelberg (1998). doi:10.1007/BFb0101003. http://dblp.uni-trier.de/rec/bibtex/conf/edbt/AgrawalGL98
4. Angluin, D.: Queries and Concept Learning. Mach. Learn. **2**(4), 319–342 (1988). http://dx.doi.org/10.1023/a:1022821128753
5. Cabasino, M.P., Darondeau, P., Fanti, M.P., Seatzu, C.: Model identification and synthesis of discrete-event systems. In: Zhou, M., Li, H.-X., Weijnen, M. (eds.) Contemporary Issues in Systems Science and Engineering. John Wiley & Sons Inc, Hoboken (2015). doi:10.1002/9781119036821.ch10
6. Cabasino, M.P., Giua, A., Seatzu, C.: Linear programming techniques for the identification of place/transition nets. In: 47th IEEE Conference on Decision and Control, CDC 2008, pp. 514–520. IEEE (2008)
7. Cook, J.E., Du, Z., Liu, C., Wolf, A.L.: Discovering models of behavior for concurrent workflows. Comput. Ind. **53**(3), 297–319 (2004)
8. Dotoli, M., Pia Fanti, M., Mangini, A.M., Ukovich, W.: Identification of the unobservable behaviour of industrial automation systems by petri nets. Control Eng. Pract. **19**(9), 958–966 (2011)
9. Estrada-Vargas, A.P., Lesage, J.J., López-Mellado, E.: A stepwise method for identification of controlled discrete manufacturing systems. Int. J. Comput. Integr. Manufact. **28**(2), 187–199 (2015)
10. Estrada-Vargas, A.P., López-Mellado, E., Lesage, J.J.: A comparative analysis of recent identification approaches for discrete-event systems. Math. Probl. Eng. **2010**, 21 (2010)
11. Giua, A., Seatzu, C.: Identification of free-labeled petri nets via integer programming. In: 44th IEEE Conference on Decision and Control, 2005 and 2005 European Control Conference, CDC-ECC 2005, pp. 7639–7644. IEEE (2005)
12. Gold, M.E.: Language identification in the limit. Inf. Control **10**(5), 447–474 (1967). http://www.isrl.uiuc.edu/~amag/langev/paper/gold67limit.html

13. Klein, S., Litz, L., Lesage, J.J., et al.: Fault detection of discrete event systems using an identification approach. In: 16th IFAC world Congress (2005)

14. Leemans, S.J.J., Fahland, D., van der Aalst, W.M.P.: Discovering block-structured process models from event logs - a constructive approach. In: Colom, J.-M., Desel, J. (eds.) PETRI NETS 2013. LNCS, vol. 7927, pp. 311–329. Springer, Heidelberg (2013). doi:10.1007/978-3-642-38697-8_17

15. Meda-Campana, M., Ramirez-Treviro, A., López-Mellado, E.: Asymptotic identification of discrete event systems. In: Proceedings of the 39th IEEE Conference on Decision and Control, vol. 3, pp. 2266–2271. IEEE (2000)

16. Meda-Campana, M., López-Mellado, E.: Identification of concurrent discrete event systems using petri nets. In: Proceedings of the 17th IMACS World Congress on Computational and Applied Mathematics, pp. 11–15 (2005)

17. Roth, M., Schneider, S., Lesage, J.J., Litz, L.: Fault detection and isolation in manufacturing systems with an identified discrete event model. Int. J. Syst. Sci. **43**(10), 1826–1841 (2012)

18. Tapia-Flores, T., López-Mellado, E., Estrada-Vargas, A.P., Lesage, J.J.: Petri net discovery of discrete event processes by computing t-invariants. In: Emerging Technology and Factory Automation (ETFA), pp. 1–8. IEEE, September 2014

19. Tapia-Flores, T., Rodríguez-Pérez, E., López-Mellado, E.: Discovering process models from incomplete event logs using conjoint occurrence classes. In: van der Aalst, W.M.P., Bergenthum, R., Carmona, J. (eds.) Algorithms & Theories for the Analysis of Event Data, vol. 1592, pp. 31–46. CEUR-WS.org, New York (2016). http://ceur-ws.org/Vol-1592/paper03.pdf

20. Wang, D., Ge, J., Hu, H., Luo, B.: A new process mining algorithm based on event type. In: 2011 IEEE Ninth International Conference on Dependable, Autonomic and Secure Computing (DASC), pp. 1144–1151. IEEE (2011)

21. Wang, D., Ge, J., Hu, H., Luo, B., Huang, L.: Discovering process models from event multiset. Expert Syst. Appl. **39**(15), 11970–11978 (2012)

22. Wen, L., van der Aalst, W.M.P., Wang, J., Sun, J.: Mining process models with non-free-choice constructs. Data Min. Knowl. Discov. **15**(2), 145–180 (2007). http://dx.doi.org/10.1007/s10618-007-0065-y

Automated Prediction of Demographic Information from Medical User Reviews

Elena Tutubalina[1] and Sergey Nikolenko[1,2,3(✉)]

[1] Kazan (Volga Region) Federal University, Kazan, Russia
tlenusik@gmail.com, sergey@logic.pdmi.ras.ru
[2] Steklov Institute of Mathematics, St. Petersburg, Russia
[3] Laboratory for Internet Studies,
NRU Higher School of Economics, St. Petersburg, Russia

Abstract. The advent of personalized medicine and wide-scale drug tests has led to the development of methods intended to automatically mine and extract information regarding drug reactions from user reviews. For medical purposes, it is often important to know demographic information on the authors of these reviews; however, existing studies usually either presuppose that this information is available or disregard the issue. We study automatic mining of demographic information from user-generated texts, comparing modern natural language processing techniques, including extensions of topic models and deep neural networks, for this problem on datasets mined from health-related web sites.

1 Introduction

Modern medical studies increasingly use nonstandard sources of information to obtain new data related to medical conditions, efficiency of drugs, their adverse effects, interactions between different drugs, and so on. One such source of information can be provided by the drug users themselves, in the form of free-text web reviews, social media posts, and other user-generated texts. These sources have been successfully used, for instance, to monitor adverse drug reactions (ADRs), making it possible to detect rare and underestimated ADRs through the users complaining about their health on social networks or specialized forums [19].

However, it may be important for the medical field to learn more than just the existence of an adverse reaction from a text review. Drugs may exhibit different behaviour on people with different age, gender, or other parameters that will often be unknown for a text scraped from an Internet forum. Hence, the problem arises to mine demographic information from free-text medical reviews.

In this work, we make the first steps in the direction of extracting demographic information from user-generated texts related to medical subjects. We have collected databases of medical reviews from health-related Web sites with user-generated content, namely *WebMD* and *AskaPatient*, and have trained models to predict the age and gender of users who wrote these reviews. We compare three types of models: baseline models with SVM and logistic regression classifiers on unigrams and bigrams, topic models with user attributes such

© Springer International Publishing AG 2017
R. Prasath and A. Gelbukh (Eds.): MIKE 2016, LNAI 10089, pp. 174–184, 2017.
DOI: 10.1007/978-3-319-58130-9_17

as PLDA and USTM, and neural models based on CNNs on top of *word2vec* embeddings. Our experiments show reasonable results on age and gender prediction, so that the resulting models can be used in conjunction with relation and information extraction to mine health-related data. We also show illustrative features and conclude that our models do indeed capture relevant information.

The paper is organized as follows. In Sect. 2, we survey related work on mining drug-related information from social media and other user-generated texts. Section 3 defines models for information extraction from text that we compare in this work, namely topic models with user attributes and a convolutional architecture for natural language processing. We present experimental results in Sect. 4 and conclude with Sect. 5.

2   Related Work

The use of social media for medical and pharmacological data mining has been on the uprising since early 2010s; the term "pharmacovigilance" has been coined for automated monitoring of social media for potentially adverse drug effects and interactions; see also media articles about these effects [8,20]. One of the first works on this subject [7] analyzed user posts regarding six drugs from a health-related social network. A comprehensive review of text mining techniques as applied to drug reaction detection can be found in [5]. We also note a Social Media Mining Shared Task Workshop (organized as part of the Pacific Symp. on Biocomputing 2016) devoted to mining pharmacological and medical information from social media, with a competition based on a published dataset [18].

In [2], authors identify ADRs from texts on health-related online forums. They used dictionary-based drug detection, extracting symptoms with a combination of dictionary-based and pattern-based methods. A lift measure (also known as pointwise mutual information) was computed to evaluate the likelihood of drug-ADR relation and chi-square test was used to evaluate the statistical significance of the lift measure. Several case studies of drugs showed that some ADRs were reported prior to FDA approval. One limitation of this work is the number of annotated examples in test data: less than 500 ADRs for evaluation. In [17], existing machine learning dictionary-based approaches were used to identify disease names from user reviews about top 180 most frequently searched medications on the forum WebMD, using a rule-based system to extract beneficial effects of the drug. In order to identify candidates for drug repurposing, authors removed known drug indications and manually reviewed the comments without FDA reports. The main limitation of this work is the lack of an annotated corpus to evaluate the proposed method. The work [22] shows an experiment for ten drugs and five ADRs to examine associations between them on texts from online healthcare communities using association mining techniques. The FDA alerts served as a gold standard to evaluate the associations discovered between drugs and ADRs. We also note a series of works specifically on Spanish language social media [9,19].

Usually, pharmacovigilance studies employ simple classifiers to extract information on drug effects or interactions. For example, to mine drug-related

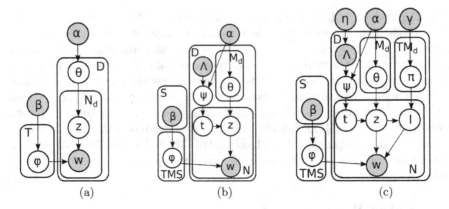

Fig. 1. (a) The basic LDA model; (b) PLDA; (c) USTM.

information from a stream of *Twitter* data, a recent work [15] uses a cascade of simple input filters followed by an SVM classifier, reporting good discovery results, while [25] proposes a weighted average ensemble of four classifiers: one based on a handmade lexicon, two on *n*-grams, and one on word embeddings.

On the other hand, drug testing and discovery of drug effects and interactions requires one to know demographic information about a user since drug effects can differ significantly depending on the user. This leads to the need to mine demographic information about the authors together with the user-generated texts themselves. When such information is provided, e.g., when the texts are collected from *facebook* users with explicitly known age and gender, there is no problem. However, in many situations user reviews for drugs and medical services are found anonymously on review web sites such as *WebMD* or *AskaPatient*; often demographic information can be known for a minority of users but not all. Hence, the problem arises to predict user demography based on the texts of user reviews. To the best of our knowledge, previous work has not attempted to automatically mine demographic information unless it was provided explicitly. In this work, we begin to fill this gap, providing first results on automated predictions of demographic based specifically on medical reviews.

3   Models

3.1   Topic Modeling

To process medical reviews, we employ the latent Dirichlet allocation (LDA) model, a classical topic model. The graphical model of LDA [1,4] is shown on Fig. 1a. We assume that a corpus of D documents contains T topics expressed by W different words. Each document $d \in D$ is modeled as a discrete distribution $\theta^{(d)}$ on the set of topics: $p(z_w = t) = \theta_{td}$, where z is a discrete variable that defines the topic of each word $w \in d$. Each topic, in turn, corresponds to a multinomial distribution on words: $p(w \mid z_j = t) = \phi_{wt}$ (here w denotes words in the vocabulary and j denotes individual instances of these words). The model introduces Dirichlet priors with parameters α for topic vectors θ, $\theta \sim \text{Dir}(\alpha)$, and β for word distributions ϕ, $\phi \sim \text{Dir}(\beta)$ (we assume the Dirichlet priors are symmetric, as they usually are). A document is generated word by word: for each word, first sample its topic index t from θ_d, $t \sim \text{Mult}(\theta_d)$, then sample the word w from ϕ_t, $w \sim \text{Mult}(\phi_t)$. We denote by $n_{w,t,d}$ the number of words w generated with topic t in document d; partial sums over such variables are denoted by asterisks, e.g., $n_{*,t,d} = \sum_w n_{w,t,d}$ is the number of all words generated with topic t in document d, $n_{w,*,*} = \sum_{t,d} n_{w,t,d}$ is the total number of times word w occurs in the corpus and so on; we denote by $\neg j$ a partial sum over "all instances except j", e.g., $n_{w,t,d}^{\neg j}$ is the number of times word w was generated by topic t in document d except position j (which may or may not contain w). In the basic LDA model, inference proceeds with *collapsed Gibbs sampling*, where θ and ϕ variables are integrated out, and z_j are iteratively resampled as follows:

$$p(z_j = t \mid \boldsymbol{z}_{-j}, \boldsymbol{w}, \alpha, \beta) \propto \frac{n_{*,t,d}^{\neg j} + \alpha}{n_{*,*,d}^{\neg j} + T\alpha} \cdot \frac{n_{w,t,*}^{\neg j} + \alpha}{n_{*,t,*}^{\neg j} + W\beta},$$

where $\boldsymbol{z}_{-j}$ denotes the set of all z values except z_j. Samples are then used to estimate model variables: $\theta_{td} = \frac{n_{w,t,d}+\alpha}{n_{w,*,d}+T\alpha}$, $\phi_{wt} = \frac{n_{w,t,*}+\beta}{n_{*,t,*}+W\beta}$.

3.2   User-Aware Topic Models

An important extension of the LDA model is presented in the recently developed User-aware Sentiment Topic Models (USTM) [23]; USTM incorporates user meta-data tags (e.g., location, gender, or age) together with topics and sentiment. In this model, each document is assigned with an observed tag or a combinations of tags, topics are generated conditioned on the document's tags, sentiment labels are generated conditioned on the (document, tag, topic) triples, and words are conditioned on the latent topics, sentiments and tags. Formally, a tag distribution ψ_d is generated for every document (with a Dirichlet prior η), for each position j a tag $a_j \sim \text{Mult}(\psi_d)$ is drawn from ψ_d, and distributions of topics, sentiments, and words are conditional on the tag a_j. The USTM graphical model is shown on Fig. 1(c). Denoting by $n_{w,k,t,m,d}$ the number of words w generated with topic t, sentiment label k, and metadata tag m in document d and extending the notation accordingly, a Gibbs sampling step proceeds as $p(z_j = t, l_j = k, a_j = m \mid \nu) \propto$

$$\frac{n_{*,*,t,m,d}^{\neg j} + \alpha}{n_{*,*,*,d}^{\neg j} + TM_d\alpha} \cdot \frac{n_{w,*,t,m,*}^{\neg j} + \beta}{n_{*,*,t,m,*}^{\neg j} + W\beta} \cdot \frac{n_{w,k,t,m,*}^{\neg j} + \beta_{wk}}{n_{*,k,t,m,*}^{\neg j} + \sum_w \beta_{wk}} \cdot \frac{n_{*,k,t,m,d}^{\neg j} + \gamma}{n_{*,*,t,m,d}^{\neg j} + S\gamma},$$

where M_d is the number of tags in document d.

For comparison, we also use a predecessor of USTM that can be though of as a simplified version of USTM, namely *Partially Labeled Topic Model* (PLDA) [16].

PLDA operates exactly the same as USTM, but without a separate latent variable for sentiment labels l_j. Its graphical model is shown on Fig. 1b, and the Gibbs sampling step proceeds as

$$p(z_j = t, a_j = m \mid \nu) \propto \frac{n^{\neg j}_{*,t,m,d} + \alpha}{n^{\neg j}_{*,*,*,d} + TM_d\alpha} \cdot \frac{n^{\neg j}_{w,t,m,*} + \beta}{n^{\neg j}_{*,t,m,*} + W\beta} \cdot \frac{n^{\neg j}_{w,t,m,*} + \beta_{wk}}{n^{\neg j}_{*,t,m,*} + \sum_w \beta_{wk}}.$$

3.3   Convolutional Neural Networks for Text Classification

The last model in our study is very different in nature from topic models. It is a classification model based on *word2vec* embeddings and convolutional neural networks (CNNs). Recent advances in distributed word representations have made it into a method of choice for modern natural language processing [3]. Distributed word representations are models that map each word occurring in the dictionary to a Euclidean space, attempting to capture semantic relationships between the words as geometric relationships in the Euclidean space. In a classical word embedding model, one first constructs a vocabulary with one-hot representations of individual words, where each word corresponds to its own dimension, and then trains representations for individual words starting from there, basically as a dimensionality reduction problem. For this purpose, researchers have usually employed a model with one hidden layer that attempts to predict the next word based on a window of several preceding words. Then representations learned at the hidden layer are taken to be the word's features.

The *word2vec* embeddings come in two flavors, both introduced in [10]: *Continuous Bag-of-Words* (CBOW) and *skip-gram*. During its learning, a CBOW model is trying to reconstruct the words from their contexts, while the skip-gram model operates inversely, predicting the context from the word. The idea of word embeddings has been applied back to language modeling in [11,12,14], and starting from the works of Mikolov et al. [10,13], word representations have been used for numerous NLP problems, including text classification, extraction of sentiment lexicons, part-of-speech tagging, syntactic parsing, and others.

In this work, for classification we use a convolutional neural network (CNN). While CNNs have been most successfully used for image processing, recent applications of CNNs to natural language processing also produce state of the art results. In an NLP task, convolutional layers are still interleaved with subsampling max-pooling layers, but this time the convolutions are one-dimensional rather than two- or three-dimensional as in images and video. Here, we use a convolutional model similar to the one recently presented in [6] for semantic sentence classification; this model has the following characteristic features:

- it is not as deep as computer vision models and involves only one convolutional layer with max-over-time pooling and a softmax output;
- regularization is achieved through dropout; the authors report a consistent and significant improvement in accuracy with dropout across all experiments;
- the model is trained on prepared *word2vec* word embeddings and does not attempt to tune word representations for better results;

– still, the authors report better results on such tasks as sentiment analysis and sentence classification than baseline techniques that include recursive autoencoders and recursive neural networks with parse trees.

4   Evaluation

For experimental evaluation, we have collected two datasets of real life health-related reviews. *WebMD*[1] is a health information services website that aims to provide objective, trustworthy, and valuable health information. We have crawled 217,485 reviews from authors tagged as "Patient" on *WebMD* and then selected 96,316 reviews about top 200 most commented drugs with a gender tag ("Male" or "Female") and one of predefined age tags: "19–24", "25–34", "35–44", "45–54", "55–64", "65–74", or "75 or over". In these 96316 documents, the total vocabulary size was 60836, the maximal sentence length was 465 words, and we found 29478 words out of this vocabulary in the *word2vec* models (most others were typos and misspellings; we did not try to correct them).

The *AskaPatient*[2] website aims to empower patients by allowing them to share and compare their medical experiences. We have crawled 113,093 reviews and selected 74,035 posts about top 200 most commented drugs with a gender tag ("Male" or "Female") and age; to unify the age prediction problem, we have split the ages into the same groups as available on *WebMD*. Since users often confuse two free-text fields about a drug, we have concatenated the "side effects" and "comments" fields, treating the result as a full review. In total, in these 74035 documents we found a vocabulary of size 63960 with maximal sentence length of 763 words, and found 33576 words out of this vocabulary in the *word2vec* models. We discarded reviews for drugs where the total number of reviews from male or female patients constituted less than 10% of the total (i.e., drugs with highly imbalanced genders) since there medications are often used to treat only female or male diseases (e.g., Mirena intrauterine or Viagra), and the gender prediction problem simply does not apply.

We report on a series of experiments with machine learning algorithms (support vector machines (CVM), logistic regression), topic models (USTM, PLDA) and convolutional neural networks (CNN). As baseline classifiers, we used LinearSVC and LogisticRegression from sklearn with default parameters (l2 for LinearSVC), TfidfVectorizer(ngram_range=(1,2), min_df=5).

As topic models, we used the Partially Labeled Topic Model (PLDA) [16] which associates each review with with a set of tags (user attributes) and the User-Aware Sentiment Topic Model [23] for modeling user meta-data, topics and sentiments for each word in a review. In preprocessing, we removed punctuation, converted word tokens to lowercase, filtered out rare words that occur less than 5 times in the each dataset. To set sentiment-based priors in USTM, we adopted the MPQA Lexicon [21] with 2718 positive and 4911 negative words. We used 20 topics, training for 500 Gibbs iterations.

---

[1] http://www.webmd.com.

[2] http://www.askapatient.com.

Table 1. Gender prediction (macro-averaged, 2 classes)

Method	WebMD				AskaPatient			
	P	R	F1	Acc	P	R	F1	Acc
PLDA	0.643	0.673	0.658	0.680	0.500	0.285	0.363	0.569
USTM word-based	0.646	0.662	0.656	0.704	0.500	0.302	0.376	0.603
SVM words+bigrams	0.677	0.637	0.656	0.752	0.545	0.555	0.550	0.640
Logistic words+bigrams	0.728	0.606	0.661	0.767	0.565	0.590	0.577	0.593
CNN 50d, rand	0.655	0.680	0.667	0.770	0.690	0.700	0.695	0.744
CNN 300d, news	0.715	0.635	0.673	0.776	0.735	0.660	0.695	0.756

Table 2. Age prediction (macro-averaged, 7 classes)

Method	WebMD				AskaPatient			
	P	R	F1	Acc	P	R	F1	Acc
PLDA	0.210	0.249	0.228	0.216	0.250	0.314	0.278	0.261
USTM word-based	0.208	0.201	0.204	0.262	0.311	0.213	0.253	0.290
SVM words+bigrams	0.281	0.246	0.262	0.297	0.312	0.243	0.273	0.320
Logistic words+bigrams	0.296	0.232	0.260	0.319	0.323	0.229	0.268	0.341
CNN 50d, rand	0.287	0.249	0.266	0.326	0.309	0.223	0.259	0.331
CNN 300d, news	0.312	0.243	0.273	0.326	0.312	0.243	0.273	0.343

Table 3. Sample topics discovered by PLDA for the tag "female" and "male".

#	Topic words
Male	
1	Muscle left pain legs hands joint neck feet pains burning arms aches body ingling walk
2	Effect sexual longer however difficult positive side sex negative control reduced libido
3	Stomach diarrhea food eat acid cramps nexium gas upset nausea reflux pains
4	Infection throat days rash itching reaction sinus body nose cough face fever
5	Meds wife make gave finally times god home big people care end rest things house stay
Female	
1	Stomach nausea diarrhea eat food cramps upset sick vomiting acid bloating gas constipation
2	Body hands rash feet reaction legs itching face swelling arms burning tingling allergic swollen
3	Days infection throat prescribed sinus sore cough headache nose antibiotic fever ear
4	Feel things happy person family dont anymore husband care longer crying depressed job

In order to get local features from a review with CNNs we have used multiple filters of different lengths [6]. We used a sliding Max Pooling window of length 2 to get features through filters. Pooled features are then fed to a fully connected feed-forward neural network (with dimension 100) to make inference, using rectified linear units as output activation. Then we apply a softmax classifier with number of outputs equal to number of classes. For all data we used dropout

Table 4. Sample topics discovered in the MedWB dataset for the tags "19–24", "25–34", "35–44", and "45–54".

Sent.	Topic words
19–24	
NEU	School dosage remember ill needed memory rate today satisfied including swollen nervous
POS	Needed intense today saved process liver teeth satisfied combined extra tolerance
NEG	Congestion nervous worrying wreck ill wondering toe distracted frightening lethargic weak
NEU	Migraines depressed mind guess fast doc wont manic takes honestly real hoping sucks
POS	Swings dreams takes doc dream single hoping approximately epilepsy wont balance hey
NEG	Mind violent beat tiny stubborn fifteen deep lying hallucinate sugars coping begging weary
25–34	
NEU	Infection seroquel methadone called experience accutane loss dizziness nauseous free
POS	Loss experience dizziness free numbness blurred returned called begin advair site important
NEG	Mind suicidal dizziness severely passed loss nauseous rough calms ineffective
NEU	Drink read bit rate sweats antibiotic issue continue gas early smoked slow avelox
POS	Wonderful pretty early benefits living satisfied advice anti-depressants eyes addicting
NEG	Anger continue daughter bit read numb paranoid downside issue eyes ran debilitating
35–44	
NEU	Pressure lyrica doc level disorder neck acne talk luck diet asked memory bottle
POS	Memory doc luck vivid talk amount everyday level diet asked moved strange pressure
NEG	Neck disorder therapy everyday ten doc itch injury strange valium mention send severely
NEU	Lexapro found prozac spasms viibryd turkey state absolutely norco easier exercise
POS	Heart experiencing exercise health reflux heartbeat effectiveness tablets aches benadryl
NEG	Found addictive aches liver crazy tolerance cure fat school changing relationship easier
45–54	
NEU	Find memory changed sex completely wellbutrin recently nightmares move understand
POS	Heart thought test energy changed checked aches understand insulin reducing patient
NEG	Upset aches fever post hydrocodone amount bowel congestion loose drowsiness
NEU	Leg immediately foot people pretty neurontin negative tests mentioned broke replacement
POS	Effective sugar abilify bit pretty regular byetta tremendously moderate intense victoza
NEG	Anger addicted seizures bit seizure negative manufacturer replacement stressful realized late

rate of 0.5, mini-batch size 50, trained for 3 epochs, and filter sizes were $[2, 3]$, each with 100 filters. For word embeddings, we tested (i) pretrained word vector *GoogleNews-vectors-negative300* (of dimension 300) and (ii) randomly initialized vectors modified during training (of dimension 50). We also tried to train word embeddings based on crawled data separately, but the results were worse and are not reported in the paper.

We performed 10-fold cross-validation on the selected data and computed precision, recall, F1-measure, and accuracy. The macro-averaged results are shown in Tables 1 and 2.

We see that the best results with respect to almost all metrics are provided by CNNs with high quality pretrained word embeddings, but CNNs without additional data or pretrained vectors come a close second, still outperforming all other models in the comparison. However, an important characteristic feature of topic models is that they can be mined for qualitative results that are easy to interpret and can validate their performance. Tables 3 and 4 show topics discovered by PLSA and USTM based on a unigram representation of reviews; for each topic, we report terms with highest weights. Note how the key terms change with age and gender, reflecting quite natural progressions.

5   Conclusion

In this work, we have presented the first results on the practically important problem of automatically learning demographic user features from his or her reviews concerning medical products or services. We have compared several different models for gender classification and age prediction: baseline classifiers that operate on words and bigrams, topic models with additional user features that can be used to predict these features, and convolutional neural networks based on *word2vec* embeddings. We have obtained reasonable results for age and gender prediction and, what is perhaps more interesting, easily interpretable topics that highlight the most common medical problems for various ages and genders.

Results of our experiments suggest that neural network models outperform topic models and baseline classifiers even in this setting, where the CNNs were not specifically tailored to this problem in any way. Hence, in further work we plan to study other neural models; for example, one interesting direction would be to go down from the level of words to the level of individual characters and try character-level models such as [24]. However, it is unclear how to get from a neural network an easily interpretable result such as a topic given by its list of words; this is also an important direction for further study.

Acknowledgements. The work is performed according to the Russian Government Program of Competitive Growth of Kazan Federal University. The work of Sergey Nikolenko was also supported by the 2016 grant "User Profiling Based on Distributed Word Representations" sponsored by Samsung. The work of Elena Tutubalina on the collection of two real-world datasets and method implementation was supported by the Russian Science Foundation grant no. 15-11-10019.

References

1. Blei, D.M., Ng, A.Y., Jordan, M.I.: Latent Dirichlet allocation. J. Mach. Learn. Res. **3**(4–5), 993–1022 (2003)
2. Feldman, R., Netzer, O., Peretz, A., Rosenfeld, B.: Utilizing text mining on online medical forums to predict label change due to adverse drug reactions. In: Proceedings of the 21th ACM SIGKDD International Conference on Knowledge Discovery and Data Mining, KDD 2015, New York, pp. 1779–1788. ACM (2015)

3. Goldberg, Y.: A primer on neural network models for natural language processing. CoRR abs/1510.00726 (2015). http://arxiv.org/abs/1510.00726

4. Griffiths, T., Steyvers, M.: Finding scientific topics. Proc. Natl. Acad. Sci. **101**(Suppl. 1), 5228–5335 (2004)

5. Karimi, S., Wang, C., Metke-Jimenez, A., Gaire, R., Paris, C.: Text and data mining techniques in adverse drug reaction detection. ACM Comput. Surv. **47**(4), 56:1–56:39 (2015)

6. Kim, Y.: Convolutional neural networks for sentence classification. arXiv preprint (2014). arXiv:1408.5882

7. Leaman, R., Wojtulewicz, L., Sullivan, R., Skariah, A., Yang, J., Gonzalez, G.: Towards internet-age pharmacovigilance: extracting adverse drug reactions from user posts to health-related social networks. In: Proceedings 2010 Workshop on Biomedical Natural Language Processing, BioNLP 2010, pp. 117–125. ACL, USA (2010)

8. Marcus, A.D.: Researchers fret as social media lift veil on drug trials. Wall Street J. (2014). http://www.wsj.com/articles/researchers-fret-as-social-media-lift-veil-on-drug-trials-1406687404

9. Martínez, P., Martínez, J.L., Segura-Bedmar, I., Moreno-Schneider, J., Luna, A., Revert, R.: Turning user generated health-related content into actionable knowledge through text analytics services. Comput. Ind. **78**, 43–56 (2016)

10. Mikolov, T., Chen, K., Corrado, G., Dean, J.: Efficient estimation of word representations in vector space. CoRR abs/1301.3781 (2013). http://arxiv.org/abs/1301.3781

11. Mikolov, T., Karafiát, M., Burget, L., Cernockỳ, J., Khudanpur, S.: Recurrent neural network based language model. Interspeech **2**, 3 (2010)

12. Mikolov, T., Kombrink, S., Burget, L., vCernockỳ, J.H., Khudanpur, S.: Extensions of recurrent neural network language model. In: 2011 IEEE International Conference on Acoustics, Speech and Signal Processing (ICASSP), pp. 5528–5531. IEEE (2011)

13. Mikolov, T., Sutskever, I., Chen, K., Corrado, G., Dean, J.: Distributed representations of words and phrases and their compositionality. CoRR abs/1310.4546 (2013). http://arxiv.org/abs/1310.4546

14. Mnih, A., Hinton, G.E.: A scalable hierarchical distributed language model. In: Advances in neural information processing systems, pp. 1081–1088 (2009)

15. Plachouras, V., Leidner, J.L., Garrow, A.G.: Quantifying self-reported adverse drug events on twitter: signal and topic analysis. In: Proceedings of the 7th 2016 International Conference on Social Media & Society, SMSociety 2016, New York, pp. 6:1–6:10. ACM (2016)

16. Ramage, D., Manning, C.D., Dumais, S.: Partially labeled topic models for interpretable text mining. In: Proceedings of the 17th ACM SIGKDD International Conference on Knowledge Discovery and Data Mining, pp. 457–465. ACM (2011)

17. Rastegar-Mojarad, M., Liu, H., Nambisan, P.: Using social media data to identify potential candidates for drug repurposing: a feasibility study. JMIR Res. Protoc. **5**(2), e121 (2016)

18. Sarker, A., Nikfarjam, A., Gonzalez, G.: Social media mining shared task workshop. In: Proceedings of Pacific Symposium on Biocomputing 2016, pp. 581–592 (2016)

19. Segura-Bedmar, I., Martínez, P., Revert, R., Moreno-Schneider, J.: Exploring spanish health social media for detecting drug effects. BMC Med. Inf. Decis. Making **15**(2), 1–9 (2015)

20. Shaywitz, D., Mammen, M.: The next killer app. The Boston Globe (2011). http://archive.boston.com/bostonglobe/editorial_opinion/oped/articles/2011/01/23/the_next_killer_app/
21. Wilson, T., Wiebe, J., Hoffmann, P.: Recognizing contextual polarity in phrase-level sentiment analysis. In: Proceedings of the Conference on Human Language Technology and Empirical Methods in Natural Language Processing, pp. 347–354. Association for Computational Linguistics (2005)
22. Yang, C.C., Yang, H., Jiang, L., Zhang, M.: Social media mining for drug safety signal detection. In: Proceedings of the 2012 International Workshop on Smart Health and Wellbeing, SHB 2012. NY, USA, pp. 33–40 (2012). http://doi.acm.org/10.1145/2389707.2389714
23. Z. Yang, A. Kotov, A.M., Lu, S.: Parametric and non-parametric user-aware sentiment topic models. In: Proceedings of the 38th ACM SIGIR (2015)
24. Zhang, X., Zhao, J., LeCun, Y.: Character-level convolutional networks for text classification. In: Proceedings of the 28th International Conference on Neural Information Processing Systems, NIPS 2015, pp. 649–657. MIT Press, Cambridge (2015). http://dl.acm.org/citation.cfm?id=2969239.2969312
25. Zhang, Z., Nie, J.Y., Zhang, X.: An ensemble method for binary classification of adverse drug reactions from social media. In: Proceedings of the Social Media Mining Shared Task Workshop at the Pacific Symposium on Biocomputing (2016)

Author Index

Printed in the United States
By Bookmasters